THE LIVING INCA TOWN

THE LIVING INCA TOWN

Tourist Encounters in the Peruvian Andes

KAROLINE GUELKE

TC TEACHING CULTURE

UNIVERSITY OF TORONTO PRESS

Toronto Buffalo London

Toronto Buffalo London
utorontopress.com
Printed and bound by CPI Group (UK) Ltd, Croydon, CR0 4YY

ISBN 978-1-4875-0810-4 (cloth)
ISBN 978-1-4875-2566-8 (paper)
ISBN 978-1-4875-3756-2 (EPUB)
ISBN 978-1-4875-3755-5 (PDF)

Library and Archives Canada Cataloguing in Publication

Title: The living Inca town : tourist encounters in the Peruvian Andes / Karoline Guelke.
Names: Guelke, Karoline, author.
Series: Teaching culture.
Description: Series statement: Teaching culture : UTP ethnographies for the classroom | Includes bibliographical references and index.
Identifiers: Canadiana (print) 2020039438X | Canadiana (ebook) 2020039441X | ISBN 9781487508104 (cloth) | ISBN 9781487525668 (paper) | ISBN 9781487537562 (EPUB) | ISBN 9781487537555 (PDF)
Subjects: LCSH: Tourism – Social aspects – Peru – Ollantaytambo. | LCSH: Tourism – Peru – Ollantaytambo. | LCSH: Tourists – Peru – Ollantaytambo. | LCSH: Tourism and art – Peru – Ollantaytambo. | LCSH: Ollantaytambo (Peru) – Social conditions. | LCSH: Ollantaytambo (Peru) – Economic conditions.
Classification: LCC G155.P5 G84 2021 | DDC 338.4/791098537 – dc23

We welcome comments and suggestions regarding any aspect of our publications – please feel free to contact us at news@utorontopress.com or visit us at utorontopress.com.

Every effort has been made to contact copyright holders; in the event of an error or omission, please notify the publisher.

University of Toronto Press acknowledges the financial assistance to its publishing program of the Canada Council for the Arts and the Ontario Arts Council, an agency of the Government of Ontario.

Canada Council for the Arts
Conseil des Arts du Canada

ONTARIO ARTS COUNCIL
CONSEIL DES ARTS DE L'ONTARIO
an Ontario government agency
un organisme du gouvernement de l'Ontario

Funded by the Government of Canada
Financé par le gouvernement du Canada
Canada

For my parents – who encouraged me to go out into the world with open eyes.

Contents

Illustrations and Tables

Illustrations

Tables

Acknowledgments

Like any journey, a book cannot be completed without the help of others along the way. Many people, not all of whom can be mentioned here, have contributed to this work. At the University of Toronto Press, I am very grateful to Carli Hansen and John Barker for believing in this project and providing support and guidance. I also thank the reviewers for thoroughly engaging with my work during challenging pandemic times; their constructive feedback made me think more deeply about the issues and strengthened this book.

My academic journey first started at Vancouver Island University, where Inge Bolin inspired my love for anthropology, introduced me to the Peruvian Andes, and provided ongoing enthusiasm and encouragement. Many years later, these experiences led to my doctoral research at the University of Victoria. There, I am deeply grateful to my supervisor Margot Wilson for sharing her experience with me over many cups of coffee; her input and ongoing support, as well as her human warmth, have been invaluable. Many thanks to Alexandrine Boudreault-Fournier for demonstrating how to push the envelope with artistic approaches and encouraging me in my own creativity. Her guidance has laid the foundation for the visual approaches included here. I am grateful to Laura Parisi for her nuanced thoughts about gender issues and Florence Babb from the University of North Carolina at Chapel Hill who provided thoughtful feedback. I further thank Cathy Rzeplinski and Jindra Bélanger for their always friendly presence in the anthropology office. During my doctoral work, I benefitted from financial support through a doctoral grant from the Social Sciences and Humanities Research Council and a fellowship from the University of Victoria's Centre for Global Studies, which also offered a lovely workspace and stimulating cross-disciplinary discussions.

Of course, this research would not have been possible without the generosity of people in Peru. *Muchísimas gracias* to Ana Caviedes Ochoa, Annushka Malpartida, and Nancy Huaman for providing a warm and supportive home in Cusco. I am also indebted to Eva Becker and Pabel Aimituma from Centro Bartolomé de las Casas for welcoming me and sharing their knowledge of tourism issues in the area. In Ollantaytambo, my thanks goes to Milagros Garcia Caviedes, Sonia Guzmán Sequeiros, Veronica Tapia Nordt, and their families for sharing their homes and stories with me. The good companionship of other new and old friends in Peru was invaluable; I think especially of Ursula de Bary, Florence Fischer, Graham Hannegan, Eugenio Aparicio Hernandez, Ana Lucia Saavedra, Gustavo Zeca, and Luisa Ditmars. As anthropologists, we utterly depend on the kindness of strangers, and I received plenty of that. I am deeply grateful to the many people in Ollantaytambo who took time out of their busy lives to sit and talk; likewise, thanks go to the many travelers who shared their experiences with me. All your stories built this work, and I sincerely hope my representations do them justice.

It is unlikely that this research and book would have been completed without good friends in my life. Thank you to Jeanne Iribarne, Bronwen Welch, Corina Schneider Fields, and Wes Stolth for great conversation, laughter, and walks in the woods. At the University of Victoria, I am grateful for my fellow travelers on the long and winding doctoral road, especially Cynthia Korpan, Betsy Hagestedt, Maral Sotoudehnia, and Marion Selfridge, and in Germany for my childhood friends Katrin and Conny Schöft who keep me grounded. A heartfelt thank you goes to my partner Trevor Moat for coming into my life, offering steady emotional support, and reading through my work with his keen eye for language. All remaining mistakes are mine. It is impossible to sum up my gratitude for my parents, Dorothea and Peter Gülke, who encouraged my explorations from early on, even though it led me far away from home.

Last, I thank my students at Camosun College and the University of Victoria for keeping me going with their questions and engagement and for reminding me of the value of the anthropological perspective.

1

Introduction

Ollantaytambo, a small community in the southern Peruvian Andes, has developed into a major stop on the tourist circuit in recent years. One morning on my way to the town's produce market, I passed by the guesthouse of Naida, a local woman of about forty years of age. We chatted for a while in the bright morning sun, as she swept her courtyard framed by Inca walls and filled with a wide array of potted plants. Soon a tour group approached through the narrow street, and an elderly man with a large sun hat peeked into the doorway, asking, "Photo? Photo?" Naida waved him in, and we stepped aside to allow the group to snap photos of her picturesque place. After they departed shouting "*Gracias, gracias* [thank you]," I asked Naida what she thought of the tourists. She responded, "Oh, they are good; they bring money."

What places have you traveled to? What sights do you like to photograph? Have you worked in tourism yourself, and what experiences have you had? In this book I analyze the interactions of local people, visitors, and tourism brokers in Ollantaytambo, located between the city of Cusco and the famous Inca site Machu Picchu. Most tourism research has focused either on tourists or on local people, but for a broader understanding, both sides of the encounter are needed

(Stronza 2001). Globally, tourism is one of the largest economic sectors and employs one in ten people (WTTC 2020). In Peru, the most visited Andean country, the number of international tourists has increased about sevenfold since 1995 (Mincetur 2020). While many travelers look for experiences different from those of home, the people living in tourist destinations are usually drawn to tourism work for material benefits, as Naida's comment indicated. Tourism can create work opportunities and has been promoted as a means of fighting poverty and increasing gender inequality; however, the costs and benefits of this development are not distributed equally. I am particularly interested in the ways gender roles and power relations are performed and challenged in daily tourist encounters. As we will see, tourism can perpetuate gendered and global inequalities, but it can also provide people with new ways to contest and renegotiate these roles. This chapter begins with a brief overview of tourism studies in anthropology and addresses a question I faced throughout my research: How are anthropologists different from tourists? Next I outline tourism development in Peru and introduce the community of Ollantaytambo, a place I first visited as an undergraduate student in the year 2000 and have seen undergo rapid changes since then. This outline is followed by an overview of my research methods with particular focus on two visual approaches, photovoice and pen and ink drawing.

ANTHROPOLOGY AND TOURISM

To begin, it is important to acknowledge that anthropology and tourism share a common history as well as an ongoing uneasy relationship (Errington and Gewertz 1989). Early anthropologists benefitted from the accounts of colonialists, missionaries, and travelers, while trying hard to differentiate themselves from these early writers of encounter narratives. Just as in the early part of the twentieth century anthropologists often excluded references to colonial impacts from their ethnographic writings, more recently they have tended to omit the presence of tourists in their study sites (Bruner 2005, 8). Yet there are many similarities between tourists and anthropologists, beginning with the fact that both are drawn by the interest of encountering "the Other," people who are regarded and framed as fundamentally different. Like tourists,

1.1 Graffiti in the center of Cusco, by Karoline Guelke

anthropologists have often objectified the people they study and over-emphasized cultural differences in their writings (Fabian 1983). These tendencies have led the discipline to a process of questioning its methods of study and representation, while tourism, as Edward Bruner points out, is still "chasing anthropology's discarded discourse, presenting cultures as functionally integrated homogenous entities outside of time, space, and history" (2005, 4). With its emphasis on play and leisure, tourism was not considered a serious topic and only became established as a subject of academic study after the advent of mass tourism in the 1970s (Gmelch 2010, 6–7). Early work was very much focused on tourism's impacts on different peoples and places. I took the photo of graffiti (figure 1.1) when walking through San Blas, one of Cusco's central neighborhoods. The controversy about tourism is clearly displayed on the wall; one can see how part of the original message was crossed out and subsequently rewritten.

This graffiti message echoes theoretical interpretations of the 1970s and 80s, when researchers viewed tourism as neocolonialism and emphasized the problems with commoditization, exploitation, and cultural loss. Tourism was criticized as objectifying and exoticizing other people and places (e.g., Greenwood 1989; Nash 1989) and even referred to as "leisure imperialism" (Crick 1989, 322). The common practice of sightseeing also came under scrutiny, most notably through John Urry's analysis of the "tourist gaze." He argues that the act of seeing is not neutral but that this gaze involves the exertion of power, which can profoundly affect those at whom it is directed (1990). However, paralleling broader theoretical debates about globalization as a heterogeneous process with complex local effects (e.g., Appadurai 1990, 1996), tourism studies also began to give greater consideration to local agency and the various ways in which people actively manage tourism development (e.g., Babb 2011; Cheong and Miller 2004; Tucker 2003). As described in the opening anecdote, Naida was not a passive recipient of the tour group but actively invited the visitors in.

Furthermore, early theorizing about tourism was characterized by structural binaries such as hosts versus guests, authentic versus staged, and everyday versus extraordinary experiences (e.g., Cohen 1988; MacCannell 1976). Tourists were generally seen as turning away from the everyday in search of more authentic, unique, and even sacred experiences. This is strongly reflected in Nelson Graburn's analysis of travel as a secular ritual, which emphasizes how the journey temporarily moves people into a liminal or transitory place where many cultural norms from home are suspended (Graburn 1989). If, while traveling, you have ever found yourself behaving quite differently than you would at home, you have experienced this liminal stage. While these approaches successfully illuminate certain aspects of tourism, they have been criticized for taking many phenomena as preexisting rather than examining how they are constructed and negotiated (Bærenholdt et al. 2004; Franklin 2004). Similarly, some have gone as far as arguing that "there is no tourist impact," as this idea erases complex historical changes and presents local people and cultures as static and passive (Castañeda 1997). This critique has resulted in the so-called performance turn in tourism studies. Rather than focusing on stable patterns and categories, this theoretical development understands tourist practices and relations as emerging through performance and negotiation (Bærenholdt et al. 2004; Uriely 2005; Urry and Larsen 2011). The concept of tourist imaginaries, "representational assemblages that interact with people's personal imaginings and that are used as meaning-making and world-shaping devices," also aligns with this approach (Salazar 2012, 864). By focusing on the processes by which local people, visitors, and tourism brokers are negotiating their roles through their interactions, my analysis in this book follows these later approaches.

PERU AND TOURISM

Tourism to Peru, particularly to Cusco* and Machu Picchu, started growing between the 1950s and 1970s; however, beginning in 1980, it declined rapidly due to the activity of the guerrilla group Sendero Luminoso (Shining Path). Following an extreme Maoist doctrine, the group was responsible for many assassinations and prompted strong counterattacks from the Peruvian military. Many indigenous people† were caught in the cross fire of these violent conflicts, and according to Peru's Truth and Reconciliation Commission, close to 70,000 people were killed in the period between 1980 and 2000. After the capture of the group's leader in 1992, the conflicts have died down but not disappeared completely (United States Institute of Peace 2018).

Like many other Latin American countries, during the 1980s and 1990s, Peru increasingly moved toward a neoliberal economic model, including financial deregulation, privatization, and trade liberalization. In the 1990s, President Alberto Fujimori began driving these developments, which were continued by his successor Alejandro

---

* There are now three main ways of spelling the city's name: Cuzco, Cusco, and Qosqo. The first is the traditional Spanish version used fairly consistently since the sixteenth century. Proponents of the *incanismo* (Inca patrimony) movement sought to highlight Inca heritage by advocating for a turn away from the more Spanish spelling, and so Cusco and Qosqo have been used since the early twentieth century as well. In this book, I am using the spelling Cusco. This version has been made official by the Cusco Provincial Council and the central government (van den Berghe and Flores Ochoa 2000, 13), and it is also the one I encountered most often during my fieldwork.

† In North America, it has become common to capitalize Indigenous Peoples in order to emphasize the political dimension and diversity. In Peru, unlike in other South American countries, there has not been a strong political movement based on indigenous identity, and few people self-identify as such (Glidden 2011). Therefore, I follow other writers about the Andean region who use the term indigenous not capitalized (i.e., Babb 2011; Canessa 2005; de la Cadena 2003; Weismantel 2001; Zorn 2004).

Toledo (Hill 2007, 437). The goal was to stimulate economic development through increased trade and foreign investment, and tourism and mining became two of the main industries (Steel 2013, 237–8). Between 2005 and 2012 Peru experienced an economic boom with an average growth rate of about 6.5 percent (Oxfam 2017). The World Bank reports that poverty rates fell steadily in that period: while in 2004 almost 50 percent of Peru's population lived on less than 5.5 US dollars per day, by 2015 this had decreased to only 22 percent (2017). However, rural areas have remained disproportionally affected by poverty. According to national statistics, in 2014 15.3 percent of the urban population lived in poverty compared to 46 percent in rural areas, with the highland regions most severely affected (Oxfam 2017).

Since the 1990s, tourism has increased rapidly. In 1995, Peru received just under 500,000 international tourists; these numbers rose to 1 million by 2002 (Index Mundi 2020) and 4.4 million by 2019 (Mincetur 2020). Records of international tourist arrivals for 2019 indicate that 57 percent come from other South American countries, mostly Chile; 20 percent from North America, mostly the United States; 15 percent from Europe; and 4 percent from Asia (Mincetur 2019, 18). About 80 percent of all tourists in Peru visit Cusco (Babb 2011, 72), and tourism infrastructure in the region has expanded accordingly. In 2019, the tourism sector's total contribution to the country's GDP was 9.3 percent, more than mining and agriculture, and it provided 7.5 percent of all employment (WTTC 2020). Tourism was predicted to continue its growth, but in early 2020 the COVID-19 pandemic brought travel to a standstill across the globe. Peruvian president Martín Vizcarra ordered a nationwide lockdown in March, closing all nonessential businesses and prohibiting domestic and international travel. Tourist numbers are lowest during the rainy season from November to March, so the shutdown coincided with the end of a low-income period, thus exacerbating the acute economic strain for those depending on tourism work. This is the case especially in the Cusco region, which lacks other major industries for employment.

Ollantaytambo, "The Living Inca Town"

The tourists taking photos of Naida's courtyard were clearly attracted by the impressive Inca walls that form the foundations of most houses in the old part of town, a feature that sets Ollantaytambo apart from other villages in the region. Ollantaytambo lies at an altitude of about 2,800 meters and is located 72 kilometers northeast of the city of Cusco (see figure 1.2). Most tourists arrive by train or bus from Cusco and then continue to the famous Inca site of Machu Picchu. The nearby town of Chinchero is currently undergoing heavy construction due to plans for a new international airport, and farmers have been resettled to clear space on the wide plain adjacent to town. In recent years, several tourist stops have been built along the road, where tourists can take photos of the mountain scenery and purchase Andean crafts.

After almost an hour's drive from Cusco, the road descends into what is called the Urubamba Valley or the Sacred Valley of the Incas. Tight switchbacks lead down to Urubamba, the valley's largest town. Here the bus crosses the river and follows it northwest along a meandering road through fields, eucalyptus trees, and small settlements of adobe and cement construction. The mountainsides are steep, and occasionally Inca terracing becomes visible. As the valley narrows and drops slightly, the first houses of Ollantaytambo come into view. The road now turns from asphalt to cobblestones and is often congested with traffic. The old part of town consists of a grid of narrow lanes; the four parallel main streets have small water canals running through, which have long provided fresh water and constant background sound to the inhabitants. (For a view of the town, see figure 1.3.) Settlement of the area likely dates back to at least 700 CE. The town and large ceremonial center were built by Inca emperor Pachacuti (or Patchacutec) in the fifteenth century, when Ollantaytambo became the residence of Inca nobility and served as civil, religious, and military quarters (Olazabal Castillo 2010, 5–6). The valley's warmer climate and fertile soil, as well as its proximity to the jungle, made the area an important resource center for the Inca empire (Hubbard 1990, 16).

In 1537, the town was the site of a major battle in which Inca ruler Manco Inca defeated the Spanish colonizers led by Hernando Pizarro. However, the victory was short-lived, and most of the land was converted to Spanish-owned haciendas (Olazabal Castillo 2010, 7). According to the Peruvian census, in 2014, Ollantaytambo had just over 11,000 inhabitants, including surrounding communities (INEI 2014). Many people speak Quechua in their homes, but due to the complexities of racial definitions in the Andes, discussed in chapter 3, the majority of people in town do not typically identify as indigenous.

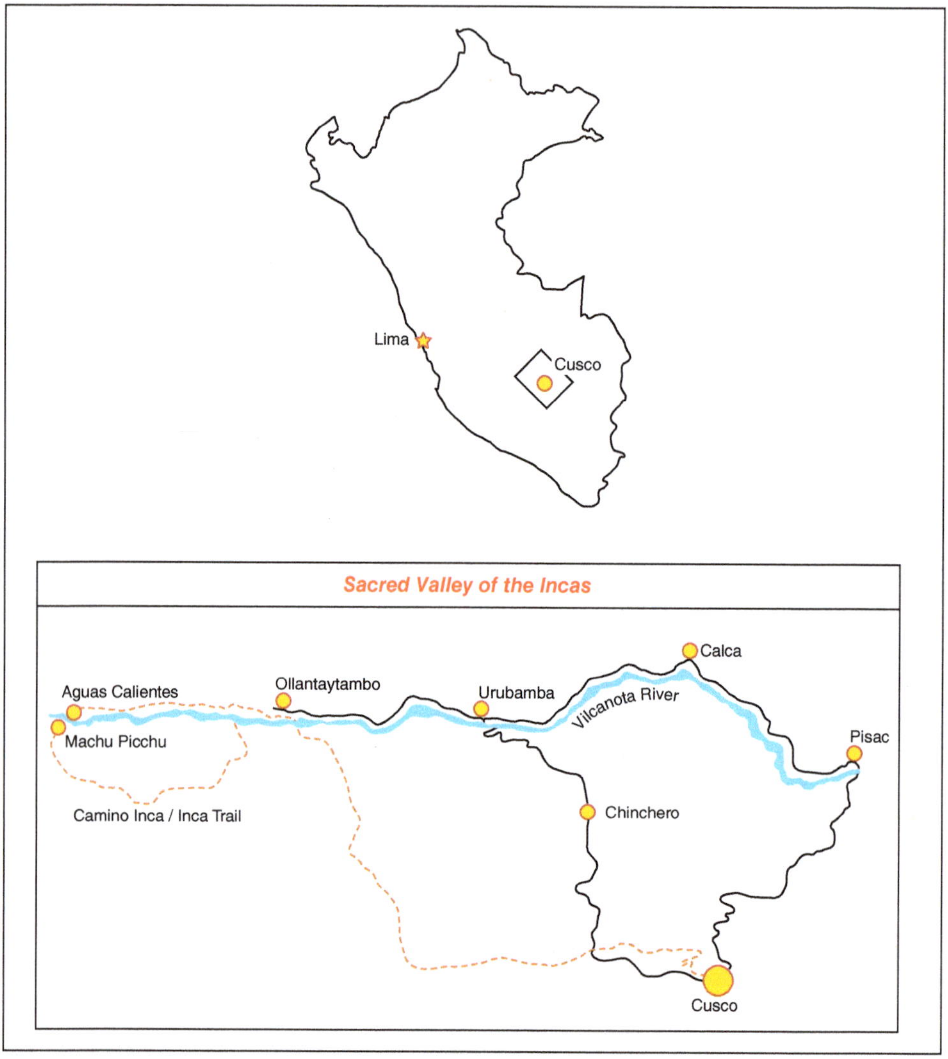

1.2 Map of Peru and the Sacred Valley of the Incas (showing the location of Ollantaytambo in relation to Cusco, Peru)

1.3 View of Ollantaytambo, by Karoline Guelke

An American traveler who spent several months living in Ollantaytambo in the late 1980s reports that there were just three small guesthouses, and he paid one dollar a day for room and board (Hubbard 1990, 38). Since then, tourism infrastructure has increased dramatically. In 2011, there were forty-seven establishments of accommodation, the majority small guesthouses with under eleven rooms (Sariego López and Moreno Melgarejo 2011, 25–6). From my informal observations, I estimate that by 2015 the number of guesthouses had increased to seventy or eighty, but apart from three larger developments, these new accommodations were still mostly small and locally owned. In 2011, forty-three restaurants were registered, again with small establishments prevailing. Most hotels and restaurants are locally owned and clustered around the two plazas and the old part of town (Sariego López and Moreno Melgarejo 2011, 26–7). Recent new developments include several large restaurants along the main road through the valley, and bigger tour groups increasingly stop there, taking business away from the smaller establishments in the valley towns.

The tourist map from the municipal office shows eighteen sites of tourist interest in the area, including the main Inca ruins, various smaller Inca and pre-Inca sites, natural attractions, and indigenous communities. The map also provides information about the town's traditional festivities, gastronomical highlights like the large white corn and roasted guinea pig, crafts, and woven textiles. A strong emphasis on Inca heritage is reflected in Ollantaytambo's official title *"la ciudad Inca viviente"* (the living Inca town). However, for the majority of visitors Ollantaytambo serves mostly as a stopover to Machu Picchu. Most tourists spend only a few hours to visit the main Inca site; they may have a coffee or a quick meal before boarding the train, while others continue on to the Inca trail, the famous four-day trek to Machu Picchu. Out of the 400,000 tourists who visited Ollantaytambo's Inca site in 2007, only 10 percent spent the night in town; the municipal government has identified this pattern as a major problem and is seeking to attract more long-term stays (Sariego López and Moreno Melgarejo 2011, 35).

RESEARCH METHODS

During my ten months in Ollantaytambo (September 2014 to June 2015), I stayed at eight different locations – one homestay, three hotels, and four small guesthouses – that all afforded me different perspectives on tourism in the community. As outlined previously, my goal was to learn about the experiences of both locals and visitors, yet these categories are very heterogeneous. For the purpose of this study, I decided to distinguish groups of participants as local people, tourists, volunteers, and foreign long-term residents. Table 1.1 provides an overview of participants and the different levels of engagement I had with them. Based on previous fieldwork and visits in 2000, 2002, and 2009, I knew two local families who provided helpful introductions to community members. As most tourists were only in town for one or two days, contact was naturally more limited. In all categories, except for the volunteers, the gender ratio is slightly skewed toward women.

Visitors or tourists are also a very heterogeneous category. According to the classic definition by Valene Smith, "a tourist is a temporarily

Table 1.1 Categories of Participants and Levels of Research Engagement (numbers in parentheses show male-to-female ratio)

Participants	Total	1 Interview, Some Participant Observation	2–4 Interviews, Participant Observation	Over 4 Interviews, Extensive Participant Observation
Locals	65 (28 m/37 f)	15 (7 m/8 f)	24 (11 m/13 f)	26 (10 m/16 f)
Tourists	62 (28 m/34 f)	41 (20 m/21 f)	21 (8 m/13 f)	nil
Volunteers	12 (6 m/6 f)	5 (2 m/3 f)	7 (4 m/3 f)	nil
Expats	11 (4 m/7 f)	2 (1 m/1 f)	4 (1 m/3 f)	5 (2 m/3 f)

leisured person who voluntarily visits a place away from home for the purpose of experiencing a change" (1989, 1). While tourists have been categorized in many different ways (e.g., Cohen 1972; Smith 1989), they are also united by shared behaviors, so that they acquire "quasi-ethnic attributes" (van den Berghe 1994, 16). In this book, I use the terms "travel" and "tourism" interchangeably. Even though many people reject being labeled as tourists, distinctions are difficult to draw and often prove meaningless to the inhabitants of host communities (Chambers 2010, 5). The majority of tourists now visit Ollantaytambo on prearranged tours, a point I will discuss later.* My focus, however, was on independent travelers who spent at least one night in town. Since they did not follow strict prearranged itineraries, I could learn more about their choices of activities in the location, and they had more time available for me. Following Erik Cohen's categorization of tourists, these were not "mass tourists" but mostly "explorers," who arranged their trips independently and associated more with local people (1972, 168). Since I was interested in examining how gender and tourism are negotiated in a cross-cultural context, my focus was mainly on independent, White, North American and European

* In 2019, 30 percent of international tourists arranged their Peru vacation as a package tour from outside the country (PROMPERÙ 2020). However, many more arrange guided tours in Cusco in order to visit the Sacred Valley.

travelers. Based on my own German-Canadian background, this selection also meant that the tourists and I shared a Western cultural background in the broadest sense.* Arguably foreign visitors have a greater impact on the Cusco area because they stand out more visually, spend more money, and rely more heavily on tourist services like restaurants, hotels, and organized tours (van den Berghe and Flores Ochoa 2000, 17). This selection seems to be in line with local views of White Westerners as the typical tourists. Commenting on a group of tourists from Barbados, one tour guide told me, "It was funny; they looked like us. They were brown, too."

Time constraints proved to be the greatest challenges in finding tourist participants. I found that the best way to establish contact was at hotels or guesthouses where I could approach during a time when travelers were not too busy, as when resting on a terrace or preparing food in a communal kitchen. Also, two local hotel owners often called me to ask for help with translation. Those situations were ideal for mutually beneficial trades: After helping with translation and providing some information about the town, I felt comfortable asking for an interview, and most tourists agreed. Often I accompanied visitors to different places, ranging from a five-minute walk to the plaza, walking around town, or sharing meals to multi-hour hikes; this type of participant observation allowed for better understanding of visitors' and locals' interactions (Fetterman 2010; van den Berghe 1994).

The numbers of volunteers and foreign long-term residents in Ollantaytambo was higher than anticipated, so I decided to include some of them in my study as well. Residents were defined as foreigners who had stayed in town for six months or more. The volunteers I spoke with spent between two weeks and six months in Ollantaytambo. Five

---

* When referring to themselves, tourists from Canada, the United States, and Europe often used the term Western or "from the West," while local Peruvians generally called them *gringos* (see chapter 3). In many disciplines the terms Global North and Global South are used to replace First World and Third World or developing versus developed countries. The goal is to "provide a more open definition of global difference, one based in social relations and cultural definitions *and* political and economic disparity" (del Casino 2009, 26). Others, however, have argued that these terms just perpetuate problematic generalizations and that more nuanced categories are needed (Eriksen 2015). By using the term Western, I follow the self-identification of many tourists. The term is not meant to imply cultural or economic homogeneity but to highlight similarities expressed in the context of tourist encounters.

worked for the Choco Museo, a small chocolate museum and store with fair-trade goods, and two for Awamaki, a prominent local NGO that sells weavings from indigenous women. Every foreign resident I interviewed had originally come as a tourist or volunteer and then decided to stay; this decision was usually related to a relationship with a local person, taking the opportunity to start a tourism business, or both. Since they had lived in town longer and were often involved in tourism through both work and travel, volunteers and residents added useful perspectives to my study.

Research participants were compensated in different ways. With local people I frequently helped with translations and assisted with computer issues and online bookings. I also taught English and German, which involved informal tutoring for two tour guides and more formal lessons in a small hospitality institute in town. In addition, I gave gifts and occasionally invited people for meals. Compensations for visitors varied; sometimes I simply answered their questions about local sites or offered small chocolates. If they agreed to a longer interview, I usually invited them for a coffee, drink, or a meal, which also allowed me to observe their interactions with local people.

Visual Methods: Photovoice and Pen and Ink Drawing

Many people expressed surprise when learning that I was including drawings in my work. Is this even scientific? Anthropology has long emphasized verbal representation, while visual methods have gained popularity only in recent decades (e.g., Banks 2005; Collier and Collier 1986; MacDougall 1998; Pink 2004, 2007). Terence Turner's work with the Kayapo in Brazil provides an interesting example of an anthropologist facilitating the use of film by indigenous people for their own purposes (1992). While film and photography have been the most common visual approaches, the methodological tool kit is expanding in different ways. Here I briefly discuss my use of photovoice and pen and ink drawing and the ways in which these methods can enhance ethnographic research and representation.

The photovoice method, a specific form of photo elicitation, was first developed for use in public health research (Wang and Burris 1997). Usually, a researcher provides individuals with cameras, instructs them on photography and ethical considerations, and asks

Table 1.2 Overview of Photovoice Participants and Images

Participants	Occupation	Photos Total	Photos Included
Edy	Runs hostel with his brother (early twenties)	7	0
Ronald	Cleans and attends at hotel, works in restaurant (early twenties)	12	1
Johnny	Paints and sells watercolor paintings on the street and in store (thirty)	22	2
Daniela	Runs hotel with partner, also sells in adjacent craft store (about thirty)	36	3
Diego	Tour guide, also helps out in restaurant his wife runs (early thirties)	37	1
Alberto	Tour guide, Diego's brother (late twenties)	42	5
Rosa	Cleans and attends at a hotel (late twenties)	70	3
Total		226	15

them to take photos of scenes that reflect their views on a specific topic. This is often followed by group discussions about the images and sometimes a public exhibit (Beh, Bruyere, and Lolosoli 2013; PhotoVoice 2020; Wang and Redwood-Jones 2001). Photovoice projects have also been used to provide community input on tourism development (Cahyanto, Pennington-Gray, and Thapa 2013; Kerstetter and Bricker 2009). Most photovoice researchers work with preexisting groups as participants, which has the advantage that people know each other and presumably are more comfortable sharing information in a group setting. While this had been my original plan as well, I found that this strategy needed to be adjusted to fit with the local cultural context. Initially, I had intended to conduct my photovoice project with students of a small hospitality institute in Ollantaytambo; however, the institute suddenly encountered problems and suspended classes. Eventually, I decided to draw on informants with whom I had already conducted interviews and felt good rapport. Two of them, Ronald and Rosa, had been students at the institute before its closure, and some of their images reflect their experiences there. I asked twelve people to take photos of what they considered particularly positive, negative, or noteworthy about tourism in their work and community, and in the end, I received photos from seven of them. Even though I had asked for between ten and

1.4 Meals prepared at local hospitality institute, by Rosa

twenty photos, people shared a highly uneven number of images. Table 1.2 provides an overview of the participants and the range of photos they contributed.

Since I had observational and interview data from all participants beforehand, I could triangulate the data and consider it in a broader context. This also helped in choosing which photos to include and gave me greater certainty that my selection reflects what is relevant to local people. Two people, Rosa and Alberto, shared the largest number of photos. As they were searching the images on their cell phones, they also commented on other tourism-related pictures and asked if I was interested. Because people had taken these photos for themselves and not for the project, they provided interesting information as well. One such example is the image Rosa took of meals she had prepared at the institute a few weeks prior. Her photo (figure 1.4) and comments reflect a positive attitude about tourism work, while other images highlight the problems with tourism.

Not all participants knew each other, and everyone had busy schedules, so I met with participants individually and asked about the photos each had taken. What became obvious in these conversations was that the images helped access more emotional and embodied experiences (Scarles 2010; Zainuddin 2009), often related to pride in their work as well as to concerns over changes in their community. In addition, though depicting "a moment in time," the photos led to conversations about broader, long-term changes (Kerstetter and Bricker 2009). Looking at her pictures of the main square, hotel manager Daniela reflected on how different the place had looked a few years earlier and how she hoped it would develop. A feminist approach should consider culture as dynamic and emergent and include people's aspirations for the future (Walter 1995, 279). Images seem successful at expanding the temporal dimension by triggering memories from the past and prompting thoughts about the future (Beh et al. 2013; Harper 2002).

Many researchers have emphasized the aspect of empowerment, arguing that photovoice can help marginalized people gain new skills and provide a means of expression and participation in decision making (e.g., Beh et al. 2013; Lykes 2006; Wang and Redwood-Jones 2001). I want to be careful with claims like these. Though disadvantaged in material terms, my participants were not marginalized members of the community. Some clearly enjoyed speaking about the images and sharing their views with me, but I do not believe this provided an exceptional avenue of expression or empowerment for them. However, the project successfully highlighted relevant issues and facilitated in-depth conversations. Importantly, it can be adapted to fit different contexts; while working with a group can be beneficial, in my case, a more individualized approach proved far more appropriate. In fact, I believe that the one-on-one interviews created a more comfortable space for participants to share information than would have been the case in a group setting. Also, I found a clear correlation between the levels of previous rapport with participants and the number of images and length of explanations they shared. Photovoice has proven to be an effective method for eliciting information and complementing more traditional ethnographic approaches, yet it is important to use it in a manner that is suited to the specific cultural context.

My second visual method consists of pen and ink drawings. Recently, a growing number of anthropologists have argued for the value of artistic approaches in ethnography (e.g., Bray 2015; Causey 2017; Pink 2007; Taussig 2011). Ingold calls for a "graphic anthropology" (2011), and Boudreault-Fournier advocates for "the role of the anthropologist as creative agent and producer" (2012). The main arguments made for drawing in the field consider the artistic merit of the product as secondary and instead emphasize how the process can enhance perception and writing, help consolidate information, and facilitate social interaction. Drawing is not simply a copying of reality but a process of selecting important elements that can focus the viewer's attention (Colloredo-Mansfeld 1993, 101). Paraphrasing John Berger, Michael Taussig writes that "a line drawn is important not for what it records so much as for what it leads you on to see" (2011, 22). We know that from photography's beginning in the nineteenth century, images have been manipulated (Winston 2005); however, even a "truthful" photograph is limited in what it can express. Photographs have "too much *excess* meaning and … too many unintended sites of connotation" (MacDougall 1998, 68). Arguably in this regard, a drawing can be more precise. David MacDougall writes that "anthropology 'makes sense' partly through elimination ... in a sense, translation is always to anthropology's advantage for it channels data through the keyhole of language, producing a condensation of meaning and leaving most of the data behind" (1998, 68). Similar to a written text, a drawing allows for this condensation, while a photograph does not. An additional point about the process of drawing is its social aspect. The camera often creates distance between people, and photography can be an aggressive act (e.g., Sontag 1977; Turton 2004). Sketching in public, however, has a "participatory dimension" (Hendrickson 2008) and can be a nonthreatening, indirect way to initiate interactions with people (Colloredo-Mansfeld 1993, 91; Ramos 2004, 149).

Most of my drawings were done after the end of fieldwork, when I worked from photographs and memory. Carol Hendrickson emphasizes how combining and alternating between written and visual recording can aid thinking through complex issues (2008, 122). While her images are produced during fieldwork, I argue that drawings done afterwards can serve a similar purpose. Rather than snapshots of particular moments, the resulting images thus reflect a deeper understanding gained over time, a process Andrew Causey calls "visual consolidation" (2017, 134). I took many photographs during fieldwork, which formed the basis of some of my drawings. One example is an image of a street vendor I saw regularly in town (figure 1.5). I took a couple of photos of her, but I wanted a simpler image, something that was simultaneously more personal and more general.

The drawings done completely from memory often depict moments when I did not have my camera with me or when it would have

1.5 Street vendor, by Karoline Guelke

been inappropriate to take a photo. In cases like these, Andrew Causey advises to take a "mental snapshot" by focusing on the important detail (2017, 135–7). I did this when witnessing two tourists taking a photo of a drunken man lying in the street and then sketched the scene later; this became the basis of a more detailed pen and ink drawing completed about a year after. The image and a description of this event are included in chapter 4. While the quick sketch was done largely as a visual recording, the later piece was part of a more complex process. At that point, I had completed my fieldwork and was engaged in the process of data analysis and writing. Returning to this scene and drawing it in more detail brought back related memories as well as the emotional dimensions of the experience (Causey 2017, 139). During my stay, I had witnessed countless instances of visitors taking photos in what I considered an invasive manner, and I had passed the same drunken man in the street many times. Thus, rather than approaching the scene as a single instance, I came to see it as emblematic. Revisiting this experience by drawing it later also brought into focus the complex ethical dimensions of the encounter and my disapproval regarding the tourists' behavior, which was combined with the awkward feeling of being implicated as well. I was writing critically about power differentials and the often problematic "tourist gaze" (Urry 1990), but what about the anthropologist's gaze? They were taking a photo; I am drawing a picture. Long after fieldwork, this visual engagement functioned as a "generative, iterative, reflective process" (Hendrickson 2008, 129). It raised issues of power and of the shared history of tourism and anthropology, and it helped me both reflect on my own role as researcher and articulate these concerns in my writing.

Some of my pieces are based neither on memory nor photos but are depictions of what I heard. One example is the image of a man walking in the dark, holding a candle and ringing a little bell. This drawing is based on local people's descriptions of the man walking around the neighborhood at night, putting a curse on an American who had recently opened a guesthouse in that part of town. The image and a discussion of this event are included in chapter 5. Andrew Causey states that "the value of a deeply experienced event is not always derived from physical perception with the eyes" but that "expanding the notion of what 'seeing' means … may open crucial doors for understanding unfamiliar cultural worlds" (2017, 148). With this drawing,

I cannot claim that "I swear I saw this," as Michael Taussig titles his book about sketching (2011), but I can offer a tentative interpretation of what I heard, which is a key task of ethnographic work.

The specific media used is also significant. My work consists of black ink lines and sketch-like watercolors that often transgress and bleed across these lines. Parts of the image remain uncolored. We recognize that ethnographic writing can only ever give us a partial view and that what we know is always embedded in unknown and ever-changing aspects of culture. According to Clifford Geertz, "cultural analysis is intrinsically incomplete. And, worse than that, the more deeply it goes the less complete it is" (1973, 29). Writing about the use of cartoons in ethnography, Dimitrios Theodossopoulos has argued that "incompleteness depicts more accurately the fluidity of social reality" (2017). The combination of pen and watercolor can effectively provide concrete detail while at the same time remaining inherently fluid and open-ended. Unlike many historical illustrations, this approach makes no claim to completeness and clearly positions the researcher/artist as part of knowledge construction. I consider pen and ink particularly well suited to reflect the open-ended process of meaning making or ethnographic approximation.

ORGANIZATION OF THE BOOK

Ollantaytambo is a small community where most people know each other. In order to protect people's anonymity, I used a number of strategies. Based on consultation with participants, I refer to people using pseudonyms; two exceptions are the local tour guide Alberto Huillca Ríos and American architect Graham Hannegan, who requested that I use their real names. In a few cases, I created composite characters by combining the traits and voices of two or more people, and sometimes I did the opposite and attributed one person's comments to two separate people. While this serves to hide people's identities, the situations and comments I describe are true in the sense that they are typical; they happened, just not always to the people indicated. Each chapter begins with a short vignette that is related to the chapter's topic and generally involves key characters who appear throughout the book. These characters and their stories constitute a condensation of traits and situations

I have witnessed repeatedly. This approach is also meant to provide some continuity; just as I came to know people over the course of my fieldwork, I hope the reader gains familiarity and understanding as we revisit these key figures in the context of different issues and themes.

Many feminist researchers have emphasized the importance of reflecting on and clearly describing our own positionality related to factors such as gender, age, nationality, and affiliation (e.g., Davalos 2008). Donna Haraway has called for clearly "situated and embodied knowledges and against various forms of unlocatable, and so irresponsible, knowledge claims" (1991, 191), and Pamela Downe writes that "we must question ourselves, our work, and our categories at the same time as we question 'others'" (1999, 141). This positioning of self extends to writing openly about our own vulnerabilities and challenges during fieldwork (Behar 1996). As a forty-year-old, divorced, and childless woman, I was an anomaly in a community where women my age typically have multiple children, if not grandchildren. This issue became especially relevant in the context of understanding gender roles and growing conflicts. In order to provide context for my findings, I add reflections about my position throughout the text.

Rapidly growing tourism development brings the promise of new experiences and access to wealth. However, people's participation and possible benefits are impacted by an intersection of ethnicity, class, and gender. My work examines how inequalities and solidarities are negotiated in day-to-day interactions. How does working in the tourist economy fit with or challenge the more traditional gender roles, and how do Peruvian women and men negotiate these changing roles? How do male and female tourists enact gender roles in a different cultural context? While I consider both men and women, my focus is slightly skewed toward the latter, since generally women have held more restricted and disadvantaged roles than men. Similar to Judith Butler, who views gender norms as resulting from active performance rather than pre-given entities (1990), I do not treat gender roles and power relations as causal factors but explore the underlying processes through which these are created, negotiated, and challenged. In a world where contact between people from different cultural backgrounds is increasingly common, understanding interactions in the context of tourism can also provide insights into cross-cultural encounters more broadly.

The focus of my book is on interactions and processes, so tourists and locals are discussed together in the context of different topics. In chapter 2, I set the scene and outline common tourist encounters. I also analyze some of the ways in which Ollantaytambo is represented in the tourist discourse and explore tourism imaginaries or common views that hosts and guests have of each other. Tourists have often been portrayed as seeking experiences of objective authenticity, but their expectations and experiences are more complex and are met by local people in different ways. Chapter 3 focuses on the ways in which tourism experiences are gendered and intersect with ethnicity. While aspects of indigenous culture have become major tourist attractions, indigenous people themselves are often objectified and marginalized. In Ollantaytambo, the increase of small-scale business opportunities appears to benefit women, and there are signs that gender roles are shifting. Also discussed are the gendered ways in which foreign tourists negotiate their travel experiences.

Tourism frequently brings together people of very different economic, cultural, and social capital, and chapter 4 provides an analysis of how local people and visitors perceive and negotiate these differences. Tourists may romanticize local people or alternatively construct them as exploitative; conversely, locals general voice appreciation for tourists while also acknowledging the immense emotional labor hosting visitors requires. It is evident that tourism perpetuates and exacerbates certain forms of inequality, yet this is often mediated by solidarity and cooperation. In chapter 5, I describe how these inequalities can result in more overt forms of conflict and aggression, including supernatural practices such as the "evil eye" and rituals performed with the intent of causing harm. I discuss means of veiled and open resistance and explore how these are expressed among community members as well as toward tourists and foreign residents. Chapter 6 focuses on spiritual and so-called romance tourism, which is prominent in the Cusco region as well. A blend of New Age tropes and Andean traditions is marketed to visitors, and many local men fashion identities to appeal to foreign women and establish relationships with them. While, on the one hand, this constitutes a selling out to foreign demands, on the other, these forms of tourism also allow local people to challenge unequal power relations, at least temporarily. Last, in my conclusion I revisit key points, report on recent changes, and offer some recommendations.

2

Tourist Encounters and Perceptions

María had grown up in Ollantaytambo; she was about forty years old with a wide smile and a thick, long braid down her back. I first met her through a mutual friend and later moved into her guesthouse in the old part of town. A wooden door opens into a small courtyard full of potted flowers and cacti, and an extended Inca building houses her family and offers five guest rooms for tourists and volunteers. During one of our many conversations in the kitchen, María told me, "You know, we used to be afraid of *gringos*.* When I was little my mother would say, 'Don't go out alone or the *gringo* will snatch you away in his backpack.'" However, usually she spoke very positively about tourists and emphasized how beneficial her guesthouse had been for her and her family. During my three-month stay with her, María made regular changes to her house. She expanded the small reception area and converted the living room into another guest room; the storage room became the new sitting area, and the adobe walls were painted white and later yellow. I joked that I never knew how I would find her place after a few days' absence. By the time I left, María had doubled the

* Across Latin America, the term *gringo* is used for foreigners, especially Americans and Europeans. It originated in Spain and often carries a disparaging connotation (*Merriam-Webster Dictionary* 2018). I have also heard it used to refer to light-skinned South Americans. Following Spanish grammatical rules, the female form is *gringa*, while the plural forms are *gringos* and *gringas* for males and females respectively.

number of beds for rent and had moved her family into a second home outside of town.

Lisa and Jonathan, two young backpackers from London, spent a month volunteering at Ollantaytambo's fair-trade chocolate store. Speaking about how much she liked the town, Lisa said, "During the day everywhere you look there are tourists, but around six o'clock you're like, 'Ah, thank God!' And it's back to a sleepy little town – everybody leaves, and it's back to the sleepy little Andean town it always was. I like that, you know, just selfishly. I don't want them to stay."

Ollantaytambo is rapidly developing as a tourism destination, and in many ways the changes in María's house reflect those in the village as a whole. Houses are rearranged and converted to hotels and restaurants; some local people move outside of the village to make space in their homes, while outsiders move in to benefit from the growing business opportunities. The community is responding to the tourist gaze (Urry 1990). Yet the majority of tourists only stay for a couple of hours, visiting the main Inca site and then carrying on to Machu Picchu; others may spend a day or two exploring the town and surrounding attractions. Like Lisa, quoted above, many visitors are seeking the experience of an authentic Inca village, whereas local people are busy adapting their town to tourism development. In this chapter, I examine "the physical and mental landscapes where the imaginaries of local residents, tourism intermediaries, and tourists meet and, occasionally, clash" (Salazar 2012, 876). Darya Maoz has proposed the useful concept of the "mutual gaze" through which both visitors and locals view and affect each other. This gaze not only implies the act of looking but more broadly considers "the ways guests and hosts view, grasp, conceptualize, understand, imagine, and construct each other" (2006, 222). In addition, examining the idea of authenticity can help us analyze tourists' expectations as well as strategies locals use to meet their visitors' needs. This chapter serves to set the scene and convey a sense of typical tourist encounters; I also introduce some of the main characters who will reappear throughout the book. Many of the themes introduced here will be explored in more detail in subsequent chapters.

OLLANTAYTAMBO: REPRESENTATIONS AND ISSUES OF AUTHENTICITY

The Moon Handbook (Wehner and del Gaudio 2011, 43) describes Ollantaytambo as follows:

> Ollantaytambo is the last town in the Sacred Valley before the Río Urubamba plunges through steep gorges toward Machu Picchu. It is the best-preserved Inca village in Peru, with its narrow alleys, street water canals, and trapezoidal doorways. The Inca temple and fortress above town is second in beauty only to Machu Picchu. In the terraced fields above town, men still use foot plows, or *chaquitacllas*, to till fields and plant potatoes. There are endless things to do in and around Ollantaytambo, which is framed by snowcapped Verónica mountain and surrounded on all sides by Inca ruins, highways, and terraces.

The writers strongly focus on the town's natural surroundings and historical features. Mentioning the foot plow, which the large majority of tourists will never see, serves to exoticize the place and locate it in the past. The Lonely Planet guide (McCarthy et al. 2013, 242) takes a similar approach:

> Dominated by two massive Inca ruins, the quaint village of Ollantaytambo (known to locals and visitors alike as Ollanta) is the best surviving example of Inca city planning, with narrow cobblestone streets that have been continuously inhabited since the 13th century. After the hordes passing through on their way to Machu Picchu die down around late morning, Ollanta is a lovely place to be. It's perfect for wandering the mazy, narrow byways, past stone buildings and babbling irrigation channels, pretending you've stepped back in time.

The presentation of destinations as part of the past, usually aimed at urban Western travelers, is common in tourism marketing (e.g., Bruner 2005; Edwards 1996, 1997), and likewise anthropologists have been guilty of describing other people as part of the past (Fabian 1983). Arjun Appadurai has pointed out that tourists often see their own imagined past in the people they encounter. Echoing modernization theory, they assume a linear progression of cultural development, in which their own culture has advanced further and others are seen through the lens of their own past (1996, 31). For Lisa, the departure of tourists in the early evening meant a return to the status quo, to "the sleepy little Andean town it always was." Reflected in this statement is the assumption that Andean culture has been static, if not stagnant, and has only been altered recently through tourism development. Also noteworthy is the description of groups of tourists passing through as "hordes." Though written for tourists, the guidebook employs a clear distinction between those addressed, the more independent backpackers, and the Others, or mass tourists. Even though Lisa had also referred to herself and her partner as tourists, in her statement quoted in the vignette that opened the chapter, tourists are clearly positioned as the Other, the ones she does not want to stay.

Attempts to differentiate themselves from other tourists have been widely documented among independent travelers and backpackers (e.g., Cheong and Miller 2004, 372; Welk 2004). One early evening, I was chatting with Joaquin, a man from Lima who ran a popular bar in town. He had just opened up for the night and was wiping the large wooden counter as the first guests arrived. Paul, a jovial Brit in his sixties, ordered a large beer and started a conversation with us: "Yeah, it's a very pleasant place. I could stay for a few more days, but it's too bad with all the tourists here." He chuckled and added, "We want to travel, but we don't want to see tourists, right?" This reflects a major paradox of traveling: on the one hand, people want to avoid other tourists, and on the other hand, they seek out familiar environments and company.

By referring to Ollantaytambo as "the sleepy little Andean town it always was," Lisa also implies the notion of a traditional and authentic culture. Debates around the concept of authenticity have been prominent in studies of tourism and have challenged the simple binary between real and fake. In tourism marketing, we commonly find a "museum-linked usage" of the concept of authenticity, referring to products and events that are produced in what is considered a traditional manner; this can be specified as objective authenticity (Wang 1999, 351). Related to this is Dean MacCannell's concept of "staged authenticity" (1976), since for something to be staged, there must be a clearly distinguishable original. In addition, Ning Wang differentiates "constructive authenticity," which results when tourists experience something as being authentic based on their own beliefs and perspectives. Regardless of the nature of the object or event, the perception of authenticity is considered socially constructed (Wang 1999, 351–2). Last, the concept of "existential authenticity" has been defined as "a special state of Being in which one is true to oneself," whether in opposition or in alignment with cultural norms (Wang 1999, 358). Edward Bruner has argued that researchers are far more concerned with issues of authenticity than tourists themselves (1991, 241). However, I found that tourists regularly referred to their hopes of finding authenticity or "the real thing," while many also questioned this concept and, as Paul's comment acknowledges, recognize the irony of being implicated.

Related to this quest for authenticity, a clear dichotomy between traditional and modern – the old and the new – is also evident in tourist discourse. Jessica, an American in her late twenties who spent three weeks volunteering at a local hotel, commented:

> As I'm taking pictures, all of them have signs saying "pizza," and all those things that are very Western, you know, in the background of this like beautiful fabric-weaving market. It's like "pizza," "Visa accepted" here, and these funny things that take away from the authentic experience. Yeah, it's an odd combination of old and new. I think that's what's weird: it's seeing the new, the commercial stuff, and then seeing like the rustic, old buildings and the water canals. Those two things just don't fit together in some way.

Jessica's expectations reflect similar imaginaries as those in the guidebooks quoted above. Her comment implies a clear distinction between the old, traditional, and authentic, which she wanted to photograph, and the new and inauthentic, which she felt was in the way. Jessica's visual approach illustrates a hermeneutic cycle of representation: existing information and images affect what tourists seek out while in turn tourists reproduce these motifs in their photographs (Albers and James 1988; Urry 1990). But based on Jessica's expectations, the elements appear incongruous: old buildings, water canals, and, interestingly, the craft market, catering specifically to tourists, are considered authentic, whereas pizza and Visa signs strike her as "funny things." The same view was expressed in the comments of an American backpacker, traveling with her partner:

> We were talking about what we wanted to experience here in Peru: the authentic culture of people who live without all that stuff, technology and that. So it's disappointing when you come into town and you see all those pizza places in the plaza and the bank machine and, well (laughs), but then again, we do like a good internet connection!

While obviously aware of the irony of her searching, she still employs the same dichotomy. This reflects the common double standard of tourists valuing innovation and change in their own societies yet seeking the stability of tradition in other places. This view was not just expressed by foreigners but also by national travelers. Luz Marina, a middle-aged dance teacher from Lima, said, "When I was here years ago, it was much quieter, more indigenous and independent, but now everything is for tourism. I think the produce market is now the only authentic part of town." While the two Americans gauged their experiences against their expectations, Luz Marina could cite past visits for comparison. Yet she, too, employed a dualistic view using the concept of objective authenticity. Implied in all these statements is also a longing for a more pristine and original state. Renato Rosaldo's much-quoted concept of "imperialist nostalgia" describes the mourning that colonial agents experienced about the people and places they fundamentally altered or destroyed. He writes that "imperialist nostalgia uses a pose of 'innocent yearning' both to capture people's imaginations and to conceal its complicity with often brutal domination" (1989, 108). In the context of tourism, this "innocent yearning" can sideline the fact that through their presence and consumption of tourist services, visitors are implicated in the very changes they lament.

LOCAL PERCEPTIONS

Whereas tourists can pretend to step back in time, as the Lonely Planet guide invites them to do, the inhabitants of this rapidly developing tourist town cannot stay locked in the past, and likely have not interest

2.1 Hotel kitchen after cleaning, by Ronald

in doing so. Like María, quoted at the beginning of this chapter, people in Ollantaytambo almost always described tourism as positive, emphasizing how the income from working in restaurants and guesthouses and from craft sales benefitted them. Women in particular mentioned that their income had allowed them to be more financially independent and better provide for their children; gendered impacts like these will be discussed in detail in chapter 3. In response to one of my standard questions about problems with tourists and tourism, most local people answered, "There aren't enough tourists." Like Rosa's photo shown in the previous chapter, other photovoice images indicated that many local people take pride in their work. Ronald, about eighteen years old and from a nearby community, included a photo of a kitchen (figure 2.1). He worked long hours in a restaurant and hotel and told me how he had cleaned the kitchen to the satisfaction of his boss the previous night. He added that he much preferred this work to agricultural labor and hoped to soon complete his training as a cook.

2.2 Garbage left by tourists, by Daniela

Several local people also commented that they learned from tourists. Alberto and Diego were two brothers who worked as tour guides. Two of the images Alberto contributed, not included here, show tourists carrying plastic bottles. He commented, "Some tourists are collecting garbage; that's really good. Here many people just toss garbage everywhere. I have learned something from tourists, and I am now teaching my son the same thing."

However, not all opinions were positive. María had mentioned being afraid of foreigners as a child, and the fear of White strangers in rural areas of the Andes has been documented elsewhere (e.g., Weismantel 2001). In Ollantaytambo, however, this fear seems to be an issue of the past, and, at present, local people mainly criticize the visitors' lack of respect. Ronald shared a photo of garbage left in a hotel room, and so did Daniela, who was managing one of the larger hotels in the main square. (See figure 2.2.) Scrolling through several images that showed trash on the floor, she shook her head and said, "This is disrespectful. We have garbage cans; why don't they use them?"

Alberto shared a photo of the town's main tourist attraction (figure 2.3), the Inca fortress and temple. He commented, "Here

2.3 Part of Ollantaytambo's main Inca site, by Alberto

is our Inca site, full of tourists. We are proud that so many people come and show an interest in our culture, in our ancestors. But sometimes they are disrespectful, too. They climb on top of the walls and do damage. Some write their names on the stones; that's very bad."

In some instances, local people also voiced concern over the fast changes in town. Naida described the physical changes in the landscape brought about by tourism and other economic developments: "There used to be just fields and trees, and the river was wider, but they just keep building houses now and making it narrower. It used to be just an open swampy area, wide and beautiful." Mauricio, a tour guide of about thirty, criticized the many new buildings in the old part of town. "Yeah, it's changing very fast," he said, "we are losing our culture, and people don't even notice." In addition to its romanticizing introduction, the Moon Handbook addresses some of these contemporary challenges as well (Wehner and del Gaudio 2011, 43):

> Ollantaytambo is also in the throes of a tremendous struggle to save its way of life against the mass forces of tourism and

2.4 Partial view of main plaza, by Rosa

> development. Trinket sellers have crowded the areas in front of the Inca temple and the train station. Nondescript pizzerias are creeping onto the main square, which is continually shaken by the passing of massive trucks bound for the Camisea pipeline in the jungle around Quillabamba.

For the first part of my stay, I witnessed the major traffic jams that often brought traffic to a standstill in the afternoons. In early 2015, the controversial bypass along the river, which had involved the destruction of some Inca terracing, was completed. While this alleviated the traffic problems somewhat, many trucks still followed the route through town, and tour-bus traffic continued unabated. Six months after the bypass opened, traffic issues still emerged as a major concern through the photovoice project. Daniela, the hotel manager mentioned previously, commented on the dangers of being hit by cars in the narrow streets. Rosa shared the photo in figure 2.4,

2.5 Electric cables in front of Inca site Pinkuylluna, by Rosa

saying that she felt that buses and cars should not park on the plaza at all.

Despite these points of criticism, the general sentiment of locals toward tourism was positive. Interestingly, I heard very different opinions from people in the city of Cusco. A friend there had invited me to her birthday party, and the conversation turned to tourism. Most people present were members of the middle class or elite, including an architect and an internationally renowned musician. The general consensus was that tourism development was destructive, and people reminisced about the good times twenty or thirty years ago when, as they said, they had Cusco to themselves. This view contrasts strongly with the opinions of people who are directly engaged in tourism work in Ollantaytambo.

Local people may also start looking at their own surroundings through the lens of the tourist gaze. This perspective is distinct from what Darya Maoz refers to as the "local gaze," which is directed back

2.6 Bicycle cart for transporting goods, by Alberto

at tourists (2006) and more reminiscent of how people internalize and adopt the gaze of the powerful as their own (Foucault 1978). While it can be expected that people working in tourism will consider their visitors' perceptions, I was still surprised by how frequently locals expressed views that were more based on tourists' concerns than their own. Several photovoice images illustrate this. Rosa contributed a photo showing the view from town up to a smaller Inca site (figure 2.5).

"Those cable are ugly", she commented, "and they are in the way when you want to take photos of the Inca sites." Rosa has lived in town for over five years and is not likely to take pictures of the Inca sites regularly, yet her comment reflects a touristic concern with a view of the old, "authentic" village unaffected by modernity. This concern may also account for her opinion, cited previously, that cars should not park in the main plaza. Alberto's photo of one of the common bicycle carts illustrates the same concern (figure 2.6).

He commented that "people shouldn't just leave the carts there; tourists complain that it spoils the view. They want to take photos of just the streets." When I questioned how people should transport their goods through the narrow lanes that cars cannot enter, Alberto conceded that the carts were necessary. Another of his photos, not included here, shows two local men stumbling in the street. He explained: "There were these drunken men; they were really loud. That's bad, especially when the tourists see it." High alcohol consumption is a problem in Ollantaytambo, but here it is judged largely based on how drunkenness may affect tourists. In the Vietnamese town Hội An, a designated World Heritage site, inhabitants have been ordered to remove or hide cables and satellite dishes in order to conform with notions of objective authenticity, which Michael Di Giovine describes as the "museumification of local cultures" (2009, 261–7). In Tiwanaku in Bolivia, state officials have suggested reverting to thatch roofs in order to increase tourist appeal, even though local people associate these with poverty and prefer metal ones (Sammells 2014). As of 2020, there are no such regulations in Ollantaytambo, though I have heard calls for this, mainly from locals disgruntled by their neighbors' building projects. However, we can see the processes of museumification not just in the perceptions of tourists but also in the ways in which local people come to view their home community.

EMOTIONAL WORK, ALIENATION, AND EXPECTATIONS

Tourism encounters involve emotional dimensions as well; many tourists hope for some form of connection with the people they meet, and hosts have to perform significant emotional labor to satisfy these demands. During the many hours spent sitting in the plaza, I could often observe the street artist Johnny at work. One of his strategies was to approach tourists and say, "Hello, my name is Picasso! Would you like to see my paintings?" This often elicited a smile from visitors and caused them to stop, but many also walked past with stony faces. Two of the photos Johnny shared with me capture the often alienating aspects of selling to tourists (see figure 2.7). The images were taken by a friend of his and show him offering his paintings outside one of the large new tourist restaurants.

2.7 Johnny trying to sell his art work

When I asked him about the difficulties of this work, he responded, "It's okay; you get used to it." Yet on another occasion, when he took me to visit the home of relatives in the country, he voiced a different opinion. Sitting in their large yard, chickens around our feet, we contemplated the impressive mountain scenery. Johnny said, "This is what I like; it's so quiet here and beautiful. I just want to sit here and paint." When I asked what he liked to paint, he answered, "The landscape, the patterns and forms I see in it. I'm so sick and tired of selling. I don't want to have to worry about money so much; money isn't everything. But I have to pay the bills." It clearly seemed that the combined challenge of material worries and the alienating aspects of selling were taking their toll.

This emotional stress was expressed by others as well. Naida and her husband Víctor had been running their small hostel for a few years, and one afternoon Víctor recounted problematic experiences with guests. In one instance, a wealthy Indo-Canadian family had arrived with seven suitcases and insisted Víctor help them carry these up from the plaza. They claimed they had booked a room with a private bathroom, while the online reservation showed otherwise. Víctor was visibly upset when he told me about this case; he said the couple had blamed him for the misunderstanding and treated him very rudely. Víctor had offered to find them another hotel, and they eventually agreed. "I just knew they'd write a bad review if they stayed," he added. "These online sites are a mixed blessing. If you get a couple of bad reviews you have to lower your prices and build up from there again." His wife Naida added, "It's exhausting; we have to be here all the time in case guests arrive, and you have to always be friendly and smile." She also recounted the story of an American guest who had accused her of having stolen his credit card. In fact, he had left it at a restaurant in Cusco the previous day, and later Víctor picked it up for him on one of his errands. "He invited us for dinner afterwards," Naida said, "but he never apologized for accusing us for no reason. I was really upset."

While one might expect differences and conflicts to be highest between Western visitors and local people, I consistently heard that the most disliked tourists were Peruvians from the coast and other Latin Americans. A local man told me that visiting Peruvian students had defaced one of the large sacred rocks in Ollantaytambo's Inca site by writing their names on it. "People from Lima are terrible," he commented, "they have no respect." María said, "Yes, our Peruvian friends – I'm sorry to say, but they are the worst. Many think we are just dirty Indians up here in the highlands." The lack of respect that many local people felt tourists showed them emerged as a major issue. This sentiment was echoed by Ade and David, an Indonesian couple, who spent three months volunteering at the chocolate museum and store. They stayed at María's guesthouse while I lived there, and we often chatted on the porch. Due to their light brown skin and dark hair, many tourists mistook them for locals, so they experienced firsthand some of the racist and classist attitudes common in Peru. Ade explained: *

> Over time, I like foreigners better than Peruvians [...] because when they come from the city, they see us here and they think we

* In quotations of people's transcribed or recorded speech, an omission made by the author is indicated by bracketed ellipsis dots, i.e., [...], and normal ellipsis markings indicate faltering or paused speech. Standard conventions apply to all other quoted material, i.e., an ellipsis (without brackets) indicates omission.

> are village people, you know, uneducated people, dirty people. Even how they call us, you know, to order – it's so rude. Especially people from Chile, Colombia, or Ecuador, especially when they are White. Because in South America they always think that the White people are better, upper class, than the indigenous people.[...] If they think that we are local, they treat us badly. Once we explain that we are not from here, that we are from Asia and working here, then they are like "ohhh" and treat us better.

Despite the lack of respect certain tourists show, many also expect some level of emotional connection with locals as part of their travel experience. Apparently, short encounters can often satisfy this need. One American backpacker in her forties described a local woman and child she had met at the bus stop. Even though she hardly spoke any Spanish, she told me that they had had a "good conversation" and that she had felt glad for "the opportunity to connect with local people." Conversely, many visitors voiced disappointment about what they perceived as a lack of emotional connection with locals, also documented in Cusco (Hill 2008, 270–1). Placing the responsibility for failed connection on members of host communities omits the issues of larger inequalities, which preserves the narrative of being a good tourist and moral being. Tom, an American volunteer, commented on the difficulty of making and trusting in these local-tourist connections:

> I can't tell often, I mean, when I talk to people it's in restaurants or stores; they are always trying to sell you something, right? I mean, the guy in the bar was very friendly, but I know he is trying to sell me something. No one seems to talk to you just for the sake of a conversation; there's usually some ulterior motive.

The frustration of being bothered by people wanting to sell is a common theme among visitors, one that I will return to in the next chapter. Jonathan, the partner of Lisa who was quoted above, summarized the difficulties of forming lasting friendships as follows:

> As a *gringo* traveling you wanna make friends, but it's hard to meet people that are willing to invest in a friendship when

> they know it's gonna be transient, you know? Because we'll go back to our life back home. We have our own network and friends, and it's a long, long way away, and the possibility of coming back to Peru is slight. And unfortunately the possibility of Peruvians coming to London is very small as well, because it's unaffordable. So, what is that friendship? That friendship you have to take for what it is, a point in time, an experience for a month.

While Jonathan and Lisa were seeking connection with locals, they also acknowledged the factors that made this difficult. Other visitors did not show this awareness. A Dutch volunteer spoke about her work with women in a nearby highland community: "I had worked with this woman; I'd stayed at her home for almost a week, and I thought we had a relationship, a friendship. But then on the last day she pulls out all her weavings and wants me to buy something, just like a tourist. That made me feel bad." I had had a similar experience. Soledad was one of the craft vendors I came to know well. I had bought small gifts at her store, and we chatted frequently. During one of the many fiestas, she had saved me a seat outside her store, where we spent most of the evening sharing beer and watching the celebration. At the end of my stay I visited her to say goodbye. "Oh, come back soon," she said. "And bring all your friends so they can buy things at my store." Like the volunteer quoted above, I felt taken aback since this overt focus on money did not fit with my understanding of friendship. Western cultures generally have taboos against talking about money, which is considered in "bad taste" or impolite. Sharing material resources happens almost exclusively within families, but far less so between friends. Both the volunteer's and my own reactions reveal a certain blindness toward our own privilege, as well as toward the different cultural norms at play. Hazel Tucker describes a similar situation in Göreme, Turkey, where tourists felt upset when a local woman invited them to see her house and then asked for money. While for the tourists economic exchange and friendship are generally considered irreconcilable, for the villagers, this dichotomy does not exist, as close personal connections and economic transactions commonly go together (2003, 123–4).

These differing views were also evident in other interactions I witnessed. For a few months, I gave German lessons to tour guide Diego, and sometimes his brother Alberto joined. We often met at a restaurant run by the brothers' aunt and uncle. One afternoon, I watched Diego enter and direct a group of six Canadians to one of the tables. They joked loudly, and a man in his fifties slapped Diego on the shoulder and repeatedly exclaimed, "Thank you, *amigo*." He then proceeded to pin a small Canadian flag to Diego's shirt, saying, "Canada, Canada." They joked back and forth for a while before Diego said goodbye and came over to my table. He told me that they had just returned from a trip to three nearby Inca sites. I had felt uncomfortable watching the scene; to me, the tourists' behavior seemed patronizing and the use of the term *amigo* contrived. But when I asked Diego about this, he responded that the Canadians had been a good group and *buenos amigos* (good friends). Like the tourists, he used the term *amigo* for people he had met that day through a commercial exchange of tourist services.

Some researchers have highlighted the alienating aspects of tourism work, stating, for example, that "tourism forces the objectification of human relations" (Bunten 2008, 384) and that inequality, intolerance, and exploitation are commonplace, while successful interactions may simply be "working misunderstandings" (van den Berghe and Flores Ochoa 2000, 7). However, both hosts and guests are also motivated to frame their encounters differently. Dean MacCannell argues that many interactions show "a certain mutual complicity ... a shared utopian vision of profit without exploitation" (1992, 28). After the commercial interaction, the Canadian gives a cheap gift, and the Peruvian leads his customers to a family business. Both make friendly gestures and seem to experience genuine positive sentiments. Here it is useful to return to the concept of existential authenticity, which is independent of toured objects or events; rather, it is a feeling of being true to oneself, which can arise in response to various tourist experiences. This can be differentiated further into intrapersonal and interpersonal authenticity (Wang 1999). In the case of the Canadian tourists, both Diego and his Canadian customer seem to have experienced a moment of interpersonal existential authenticity, and, as addressed above, for Diego, the combination of economic exchange and friendliness or friendship may have been a much more familiar one than for the Canadian.

Knowledge, Power, and Tourism Brokers

As some of this chapter's examples illustrate, local people return the gaze and can exercise power in different ways. As Michel Foucault discusses, power and knowledge are closely intertwined (1980). While visitors usually have limited knowledge of a site and its people, locals working in tourism develop a keen understanding of tourists and become expert judges of their appearances and behaviors; this acquired knowledge can function as a "balancing mechanism" between wealthier tourists and local people in developing countries (MacCannell 1992, 31). Indeed, many of the tourism workers I spoke with commented that they had learned to judge their visitors quickly. Diego described the expertise that he and his brother Alberto had developed:

> We know the different types of tourists; some are problematic, some are fun, some ask a lot of questions. We can tell quite quickly which type they are, and then we just go with that. For example, if it's a serious person, you don't make many jokes. If it's a fun person that you like, well, then you try to make them laugh. And there are people who are a bit special, right? I can tell quite well, just by looking at them. And I know who is German, American, Dutch, Japanese, or Canadian – just by looking.

Knowing how to judge visitors allows tourism workers to relate better to their customers and potentially to make more money. When I confronted a taxi driver who had overcharged me, he laughed and said he usually asked a bit more from tourists. "I can usually tell if people are able to give a little more," he said. Judging tourists can also be useful in other ways. Several times, as I was sitting with Naida at her guesthouse, I witnessed her send away people asking for rooms, claiming that she was fully booked. Usually the people turned away were "hippie type" traveling vendors and Latin Americans. When I asked Naida about this, she replied that she could tell when guests were trouble and that she preferred to leave rooms empty to keep the peace. Local people use many different strategies of resistance, which I will discuss more in chapter 5.

It is important to recognize that tourism does not only involve hosts and guests but also agents, guides, and other tourism brokers who affect and direct tourist encounters. The tourist gaze is, in part, constructed by guides and brokers, who direct what is seen and what is not, resulting in more complex flows of power (Cheong and Miller 2004, 383). David, one of the two Indonesian men introduced above, observed how strictly tourists can be controlled by their tour guides.

> Sometimes we get tourists in the shop, and I ask them, "Have you explored the old houses up there?" I always try to promote that because they are such beautiful houses. It's amazing to me, but most people just go see the ruins.[...] In the bus you just see a little bit, and the guides yell, "*Vamos, vamos* [let's go, let's go]." You know, like the animals: *vamos, vamos*. They pay so much for the guides, and then they miss a lot. Sometimes the guides drop the tourists in the train station, and their train leaves like three hours later, but they just spend the time there doing nothing. The guides just hurry them on; I've seen it many times. Why *vamos, vamos* when there is time?

One morning, I was standing in the street chatting with Anja, a longtime German resident in her early twenties, when a tour group appeared. The guide briefly stopped to point out the trapezoidal Inca doorway and then urged the group to move on. "Like sheep," Anja commented. "And they never get to see how beautiful it is just a little ways up the valley." Importantly, directing the tourist gaze also affects who gains economically. Lisa commented on her observations from work at the chocolate store:

> It gets a bit demoralizing sometimes, because people are rushing; they literally have to get on the bus right away, and I think the guides have a lot of power. Because if they have a sort of relationship with a business, and they take the tourists there ... like one day a guide came in and yeah, we made so much money. He brought a group of like fifteen people, and he was talking it up a bit, you know, like "Guys, this is the organic chocolate shop." And so we made a lot of money.

Tour groups usually stop in the craft market below the main Inca site; sometimes, tour guides will direct their customers to different stalls and tell them where to buy. This can make a crucial difference for vendors. One woman has an arrangement with a guide and will guard his group's bags while they visit the ruins; in exchange, she receives an endorsement for her wares, which usually translates into good sales. Another local woman sells crafts and weavings out of her courtyard and invites people in to see her Inca home, which is decorated rather theatrically with corncobs, Inca tools, and a stuffed condor. The many guinea pigs roaming about add tourist appeal. When I asked her how this work was going for her, she complained that not many tourists find her place. "But it's much better now that the guides started bringing in groups." She added that she often had to rely on the guides to translate because she spoke only a few words of English. The lack of English skills, common among most vendors I spoke with, is an important factor that puts local people at a disadvantage. Several people also complained that tour guides give outright false information. Adriana, a local woman in her late twenties, said, "Often they just make stuff up as they like, but it's not true. I've heard a guide say that the parts that stick out of the rocks ... that the Inca used to hang their hats on them – their hats! They didn't have hats like we do!"

As addressed in the introductory chapter, tourists are a highly heterogeneous group. Some backpackers fund their trips by selling self-made goods like jewelry; thus they occupy an interesting position in tourist interactions in that they often sell their goods to other tourists, thereby taking the role of, and sometimes competing with, local people. To avoid competition, some of them produce work that is sufficiently different, such as a Spaniard selling small wood carvings and a Chilean making wallets out of unusual materials, such as milk cartons and Jehovah's Witnesses' brochures. Erik Cohen would have categorized them as "drifters" (1972, 168). Local people were usually very critical of traveling vendors and associated them with drug use; both María and Naida cautioned me about spending time with them. Studies indicate that backpackers tend to spread money into more remote areas and support more small-scale local businesses (Cohen 1972; Maoz 2006, 223). However, despite trying to differentiate themselves from tourists, they often share many attitudes with them. Sabrina and Rafael were traveling vendors from Brazil. One afternoon, I spent an hour sitting on the pavement of the plaza with them; their brass jewelry was laid out on a small cloth in front of them, and while working on other pieces and sharing a bottle of beer, they called out to passersby. Rafael told me about their recent highland trek:

> It was amazingly beautiful, the mountains all green and white with the snow, and once in a while the local people dressed in bright red. They seem to have a simpler and more spiritual life. We came across these four women, sitting just way up there in the mountains, all dressed in their handwoven clothes and with those hats [...] and they just sat there and, I don't know, contemplated things.

In his analysis of images in tourist brochures, Graham Dann labels one of the main categories "natives as scenery." These are photos that show major landmarks or attractive landscapes and include local people in traditional dress. He argues that people "are displayed as stage extras, artists' models, objects which have replaced people, to be gazed at with impunity" which serves "to add a touch of local colour" (1996, 69–70). Similarly, Rafael's comment reflects a view of locals as part of the landscape; among the mountain scenery, the natives are reduced to providing colorful elements of red. As opposed to the hikers, locals are seen as passive: "they just sat there." Also, his reference to their lives as "simpler and more spiritual" is a common theme, which I will explore later.

Yet Sabrina and Rafael also had very different interactions with local people. Earlier that day, I had observed two young women in the traditional dress of the nearby highland communities stop and examine the couple's jewelry. Rafael held up different pieces, and the women laughed and exchanged a few words with him before moving on. I was struck by the image that in many ways constituted a reversal of the standard tourist encounter between the "colorful locals" and foreign visitors. Rather than local people offering goods or posing for the tourist gaze, in this interaction they had become the potential customers. I did not photograph this brief interaction but chose to draw it later (figure 2.8). The fact that this scene surprised me helped me recognize how much I had become accustomed to seeing foreigners consuming all that is local, rather than vice versa.

When I later asked about this interaction, Rafael said that they had learned a few Quechua phrases and tried to use them as much

2.8 Brazilian traveling vendor, by Karoline Guelke

as possible. Pointing to the intricately woven belt he was wearing, he added, "We also barter a lot. This here – I got it from a woman up in the highlands, and I traded her for a bracelet." Like other backpackers I spoke with, they had made an effort to connect more with local people, yet at the same time displayed some of the general stereotypes foreign visitors often hold.

The power of outsiders was further illustrated by another case, which also brings us back to the issue of authenticity. During the time of my stay, I often saw a man in full Inca costume and wig standing by the entrance of the main Inca site; I refer to him here as "the Inca," which was the term local people used for him. Usually, I would see him joking with tourists and inviting them to take a photo in exchange for money. (See figure 2.9.) His regalia as well as his behavior were theatrical – simply an example of badly "staged authenticity" (MacCannell 1976)?

During an interview, "the Inca" told me he was from Peru's north coast and had come to the highlands for the opportunities in tourism work. He mentioned that sometimes, referring to the Inca site, he greeted tourists with "Welcome to my friggin' house." With a big grin he added, "It works; I'm making good money." Tourists I spoke with mentioned him frequently. While a few rolled their eyes and called him fake, most of them laughed and made comments such as "he's just great" or "that crazy guy was a lot of fun." Erik Cohen refers to "postmodern tourists" as those who are not striving for experiences of objective authenticity but find enjoyment in a variety of touristic events and experiences, including those that could be considered inauthentic (1995). This is what "the Inca" caters to. He presents himself to the tourist gaze, but by clearly signaling that his performance is a parody, he produces a lighthearted mockery of the quest for authenticity, which often succeeds in delighting tourists and putting them at ease. He is not staging authenticity but actively playing with it, and the entertaining and jovial interactions he provides often result in an experience of "existential authenticity" for tourists (Wang 1999). This act can also be seen as a form of self-commodification, an expression of personal agency that can help people avoid some of the alienation of serving tourists (Bunten 2008, 381).

However, while some local people expressed mild amusement, most did not respond favorably to "the Inca's" approach. Several of

2.9 "The Inca" posing with the author, by Margot Wilson

them pointed out that he was not from the area and that he misrepresented the Inca. An experienced tour guide from Cusco said, "He is basically begging for money. The Incas were our ancestors; they were rich and noble. They were not beggars. I don't agree with what he is doing; he is presenting a false image." Inca culture and identity are important features of the Cusco region and constitute prominent tourist attractions. However, as the case of "the Inca" shows, an outsider can quite easily appropriate this heritage and present and market it effectively, even in ways that contradict local views.

From a strategic Inca stronghold to an international tourist destination, Ollantaytambo was never just a "sleepy little Andean town," yet there is no doubt that it is now in the midst of rapid economic development and cultural change. The community has become a site onto which visitors project their longings and imaginaries; some may be disappointed in their search for objective authenticity but find experiences of playfully staged and existential authenticity instead. Tourists tend to be more concerned with their own experiences than with the effects of their presence on locals (e.g., Noy 2004). Like María rearranging her house, the community adjusts in order to benefit from the growing tourism business. Many local people expressed pride in their work and spoke positively about tourism and the opportunities it brings; however, some find the lack of respect seen in some tourist behavior to be concerning. Guides, often from outside the community, hold substantial power in representing the town and directing the tourist gaze. The daily encounters between locals, tourists, and brokers indicate that the gaze is mutual and that power can flow in complex ways. These power relations are also affected by gender and ethnicity, which I discuss in the next chapter.

3

Negotiating Gender and Ethnicity

I first met Diego on the steep hill of Pinkuylluna, the smaller of the two Inca sites overlooking Ollantaytambo. I had walked up in the morning and was sketching the mountainside across the valley when a man in reflective sunglasses and baseball cap approached and offered me a guided tour. I declined, but we began a conversation. Diego was about thirty years old, married, and had two young sons. He was fluent in English, spoke basic French, and wanted to learn German as well, so I agreed to give him language lessons in exchange for interview time. Diego had been working in tourism with his younger brother, Alberto, for years, and his wife, Ana, was working long days at one of the town's larger hotels. Several times I saw him running through the streets, and he would tell me, "I have to go home to make dinner for the kids. Ana is still at work."

Toward the end of my fieldwork, Diego mentioned that Ana was now working in the small restaurant they had opened out of a converted room of their house. He invited me to come by, so one afternoon I walked up the cobblestone street until I saw their newly painted sign. The upper part of town, where the small lanes narrow even further, is mainly residential; only a couple of small shops, a bakery, and women selling homemade corn beer cater to locals. Diego proudly showed me how he had decorated the restaurant: bright orange walls with colorful wool hats and blankets typical of the region, flowers on the wooden tables, and a new TV in the corner. As Ana was preparing tea for us, she occasionally checked on their two sons playing in a room adjacent to the kitchen. Diego smiled and said, "This is good. Ana can work from home and keep an eye on the children; it's much better this way. And if everything goes well, we'll add a room or two to rent out."

Tourism and gender intersect in multiple ways. Existing gender roles affect what types of tourism work are accessible to women and men, while, in turn, the impact of tourism changes how these roles are defined (Kinnaird and Hall 1996). Ana's work at the hotel had mainly consisted of cleaning rooms while in her new restaurant, she prepared and served meals, activities that fit with the traditional gendered division of labor. Yet she and Diego were also renegotiating and challenging these roles. Statistics show that globally slightly more women work in the tourism sector than men, and the pay gap is slightly less than in the broader economy (UNWTO 2019). Gender affects the experience of travelers as well. Male and female tourists' behaviors and views may clash with those of the cultures they visit, and most female travelers feel the need to adjust their behavior in order to minimize risk. With a somewhat greater emphasis on women, this chapter examines the main ways in which tourism experiences are gendered. While I did not question people specifically on their gender identities or sexual preferences, a few participants chose to comment on this topic, so some experiences of LGBTQ+ travelers are included here. In general, however, people's comments seemed to reflect normative notions of masculinity and femininity. My findings indicate that, mostly through the increase of flexible, small-scale business opportunities, tourism can offer significant benefits to women in Ollantaytambo. However, while aspects of indigenous culture have become major tourist attractions, indigenous people themselves often remain marginalized. As in the city of Cusco, we find women and children posing for photos, but the members of highland communities are finding ways to negotiate their participation in tourism in more equitable and communal ways. Last, I consider some of the ways in which tourists negotiate their roles, or, in other words, how they perform gender in the context of travel.

RACE, ETHNICITY, AND GENDER

As feminist scholars remind us, it is important to acknowledge the intersectionality of different aspects of identity, such as class, gender, and ethnicity (e.g., McCall 2005; Shields 2008), so this is where I begin. Local Andean culture is one of the big tourist attractions of the region, and images of indigenous people and cultural events are ubiquitous in tourism advertising. Yet who is defined as indigenous is a complex issue. I start with a brief overview of racial constructs in the Andes and the way these intersect with gender. Based on their colonial history, populations across the Andes are characterized by both Spanish and indigenous ancestry. Many ethnic and racial terms exist, but people are usually divided into three broad categories: White; *mestizo*, referring to a person of both European and indigenous origin; and "Indian" or indigenous.* A further distinction, *cholo*, can describe a person whose ethnic characteristics fall somewhere between *mestizo* and indigenous (Mitchell 2006, 50–3; Weismantel 1988, 34). What is important to note is that these categories are highly variable and situational; definitions vary not only from region to region but also between different parts of the population and even individuals (Canessa 2005; de la Cadena 2003; Weismantel 2001). Distinctions are based less on phenotypic differences, which are not very prominent, than on cultural and socio-economic markers, such as education, occupation, and clothing, and map closely onto social class (Colloredo-Mansfeld 1998, 1999; Mitchell 2006, 53; Weismantel 2001). This means that, for example, a woman who lives as a subsistence farmer in a rural community, speaks mainly Quechua, and wears homespun clothing will be regarded as indigenous, yet, if she moves into a larger town, learns Spanish, and wears mainly store-bought clothes, she is likely to be thought of as a *chola* or *mestiza*.† Thus, instead of saying that most of the urban population is *mestizo*, Andean people view the very acts of living in town, speaking Spanish, and being part of the cash economy as conferring *mestizo*

* Using the concepts black and white in reference to humans can falsely naturalize these categories; by capitalizing the term *White*, I intend to indicate that, just like *Black*, it is a culturally constructed category and not based on physical reality.

† I am following the Spanish rule of using the terms *chola* and *mestiza* when referring to women and *cholo* or *mestizo* when referring to men or the population as a whole.

identity. People use the terms *indio* (Indian) and *cholo* as insults for others regarded as lower on the social scale, and rarely does anyone self-identify as *indio* (de la Cadena 1995). While the concept of "race" as a clear physical distinction is fundamentally fictitious, it remains a powerful social fact that fuels different forms of discrimination (Colloredo-Mansfeld 1999; Orlove 1998; Weismantel 1988, 2001).

Running counter to this, we also find a different process. In many parts of the Andes that were formerly part of the Inca state, the identification with the Inca past is strong; this is especially true for Cusco, the former Inca capital, and the surrounding area. Peruvian *indigenismo*, sometimes termed *incanismo*, is a movement that developed in the nineteenth and twentieth centuries as a protest against Spanish domination and still continues today. It focuses on the glories of the Inca past and depicts the Inca state as a benevolent, nonexploitative model empire (Arellano 2004; Barrig 2006; van den Berghe and Flores Ochoa 2000). Many of the concerns of *incanismo*, such as the preservation of Inca architecture and the promotion of indigenous culture, converge with the interests of tourists and have been turned into marketable commodities. *Incanismo* has been primarily directed by urban elites, who consider themselves legitimate heirs to the Inca while often distancing themselves from contemporary indigenous people, whom they view as "degenerate, ignorant, [and] backward" (van den Berghe and Flores Ochoa 2000, 12).

Most people in Cusco will point to the countryside and refer to everyone outside of the city bounds as indigenous. However, due to the long history of racism and discrimination, most inhabitants of Ollantaytambo reject this label and use the term *mestizo*. On a couple of occasions, I heard people refer to themselves as *cholos*, which carries a negative connotation, or as *serranos* (highlanders), a more neutral term. Following the definitions of people in Ollantaytambo, I use the term indigenous or Quechua when referring to people from the smaller highland communities where the Quechua language is still spoken. Gender roles vary in the different segments of the populations, and interpretations are contested. I will briefly outline two patterns that map roughly onto *cholo/mestizo* versus Quechua/indigenous populations and can be seen as ends of a continuum. While simplifying much greater complexity, these generalizations are meant to provide a starting point for discussing how gender, ethnicity, and tourism intersect differently for people in Ollantaytambo and their highland neighbors.

In most areas where the Spanish have dominated for the past five centuries, we now see gender relations strongly modeled on Spanish culture (Weismantel 2001, 49). This applies to large parts of the *mestizo* population that mainly lives in small towns across the Andes. Women's work is centered on household chores, food preparation, and childcare, while men work in various positions outside the home. As opposed to the Inca system of parallel hierarchies, where the different roles were likely recognized equally (Silverblatt 1987), in the *mestizo* system, private work is typically valued less than public duties. Fitting with the gendered divisions into public and private spheres are the complex ideologies of *machismo* and *marianismo* common across Latin America (Duran 2001, 140). These constructs vary for different parts of the population and are contested in numerous ways (Gutman 1998), yet we can identify some general characteristics. *Machismo* constructs men as superior, dominant, and aggressive. This notion has been related to colonialism: through the brutal domination of the Spanish, local men were rendered powerless to protect themselves and their families, so their aggressive masculinity can be seen as a reaction to this history as well as to the continued subjugation from dominant elites they face today (Garcia 1997; Mirandé and Enríquez 1979). Conversely, through *marianismo,* women are viewed as saintly, passive, and inherently connected to the domestic sphere, a construct based on the perceived characteristics of the Virgin Mary (Navarro 2002; Stevens 1973). While men are clearly accorded the dominant position in this system, women can exercise indirect influence and thus affect the decisions men execute in the public realm (Bourque and Warren 1981, 52; Weismantel 1988).

Gender roles in Quechua communities have been interpreted in different ways. Some researchers argue that gender relations in indigenous societies were asymmetrical even before the Spanish conquest (Barrig 2006, 112) and that "rural Andean society is heavily marked by sexual hierarchy" (Bourque and Warren 1981, 4). I frequently heard urban Peruvians and foreign residents describe Quechua women as shy and subordinate. However, others emphasize the egalitarian elements in these highland communities. Both men and women take part in agricultural work; they have distinct yet complementary duties (e.g., Allen 1988; Bolin 1998, 2006; Canessa 2005; Zorn 2004). There are no strong taboos against performing the other gender's work (Bolin 2006, 53), which contrasts strongly with the situation in *mestizo* society where women's occupations are valued less, so by performing these tasks, men risk a loss of status. Local gender ideologies define men's and women's tasks as roughly equally important, and women maintain control of important resources such as land, animals, and food.

The system of gender relations in Ollantaytambo falls roughly between these two models. In general, I saw less pronounced expressions of *machismo* and *marianismo* than reported elsewhere, though the picture varied. Diego commented:

> Yes, *machismo* exists here. Many men are very … the woman has to be in the house, cooking, looking after the children, while he can go out with his friends drinking. He doesn't want her to go out. Sure, women go out, too, but sometimes when she goes out with her friends, her husband may cause a fight; he may hit her.

His description indicates the greater power position of men as well as restrictive domestic responsibilities for women. As Florence Babb describes for market vendors in the Peruvian town of Huaraz, local women who work outside the home tend to perform the typical double shift of paid work plus domestic chores, resulting in long work hours (1989, 141–3). Many foreigners also mentioned the gendered division of labor they witnessed, commenting on the women's heavy workload. Ade and David, the Indonesian volunteers, described their experience at the chocolate store, where, to the amazement of their female coworkers, they often prepared their own meals in the workshop kitchen. As Ade said,

> We clean, we wash the dishes, we never feel ashamed to do any of that.... You know, it's dirty – we clean it. And the local girls here ... it's like something really strange for them, you know, there are men cooking! I think that's a gender difference. For them cleaning, cooking is for women, and men is farming, and the man will *never* clean the house, will *never* wash the dishes, will *never* do the cooking. They sit there and stare at us when we cook, like wow, how do you know how to do this? And you can see it in their eyes: hey, I want to be your wife because you can cook! Those women, they worship us!

This fits with my own observations. While I witnessed a few men helping with cooking and housework, most household chores were performed by women. However, women have ways to exercise power both directly and indirectly. When asked about the topic of gender, María said, "The men in Peru are *machos*, especially here in the highlands. They do what their mothers say." She proceeded to tell me about her husband who, when they first started dating, only listened to his mother. This intertwining of patriarchy and the power of older women has been documented in other parts of the Andes (Bourque and Warren 1981; Weismantel 1988). Women also have more direct avenues of influence by controlling the household resources, a point confirmed through personal conversation with Elisabeth Kuon Arce, an anthropologist with many years of experience in the Cusco region. In the families I came to know, I often witnessed women making decisions about household spending. However, based on internal power relations and gender ideologies, resources may not be evenly distributed within households, and men may withhold their income from other family members (Benería 2003, 35; Costa and Silva 2010). Kate, a British resident running one of the town's larger hotels, told me that women tended to look after the family's money. In her work supporting porters on the Inca trail, she had witnessed that many times the men's wives would come to collect their wages so that their husbands would not spend it on alcohol. While women in Ollantaytambo usually find ways to control household spending, resources can certainly be contested within the household.

Sanctions against women in public are not nearly as evident in Ollantaytambo as reported elsewhere. The vast majority of market vendors are women and so are most shoppers. Neither do there seem to be strong sentiments against informal and jovial contact between unrelated women and men. Street vendor Johnny frequently teased a female colleague, Tina, who reciprocated vehemently and often hit him playfully. Adriana, whose family I stayed with for a while, spoke one evening about her boss at a small hotel: "One time he grabbed my arm; it was jokingly, but I told him, 'Don't mess with me, or I will tell my mother' – and he said, 'No, no, please don't tell your mother; I will leave you alone.'" Her mother, Valentina, commented that such joking was inappropriate. Interestingly, she seemed less concerned with the man's actions than with her daughter's response, reflecting a generational gap in what are considered appropriate interactions between the genders and possibly a greater acceptance of violent behavior from men. Also noteworthy is that Adriana's threat consisted of telling her mother, not her father or two brothers. Even though it was meant largely in jest, this comment points to women's power and influence in the community.

Along with *machismo* we typically find strongly heteronormative views. Adriana commented that she had seen a gay couple being verbally harassed and, on another occasion, a young man beaten for being gay. When the issue came up in conversation, comments by locals were often disparaging. In general, gender roles in Ollantaytambo show elements of both indigenous and *mestizo* communities, and they are also currently undergoing significant change. Adriana commented that "women's roles have really changed in the last five to seven years. Women have emancipated themselves; they don't let themselves be dominated by men. Sure, there are always cases of violence in the home, but it has changed a lot."* María concurred; she attributed these changes largely to development programs and to new work opportunities in tourism, which I will consider in the next section.

GENDER AND TOURISM

For both local people and visitors, the experience of tourism is gendered. The gender norms of host and guest societies impact whether and how people can benefit from tourism, and a better understanding of these processes is needed in order to advance gender equity in tourism development (Ferguson 2010b; Uriely 2005). Based on nearly all measurements of human rights and human development, globally, women still fare worse than men (Parisi 2013, 439; Rai 2011a, 20), and tourism has been promoted as one strategy of fostering economic development and greater equality (Ferguson 2010a; UNWTO 2019). However, rather than relying on generalized models of universal subordination, one must investigate the processes that bring about women's subordination in a specific context (Mohanty 1997, 84–5).

* According to a national survey conducted in 2017, over 65 percent of Peruvian women have experienced physical and/or sexual violence from their partners over their lifetimes. Percentages are higher in the highland areas, and overall rates have dropped slightly across the country over the past decade (Observatorio Nacional de la Violencia contra las Mujeres y los Integrantes del Grupo Familiar 2020).

This approach also parallels tourism studies' turn away from static categories. In addition, it is important to consider how gender intersects with factors such as class and ethnicity (McCall 2005; Shields 2008) and how it is shaped by colonial history (e.g., Lugones 2010).

Research into women's roles has shown that women can maintain their existing influence and increase their status if, first, they have control over productive resources, such as land or animals, or the opportunity to earn wages and control their income, and, second, if existing gender ideologies allow or even support women's participation in production and decision making (Friedl 1991). However, often women's work in tourism is limited to an extension of their previous roles, such as domestic chores like cleaning and cooking (e.g., Feng 2013; Ferguson 2010b; Sinclair 1997). Also, economic growth alone does not automatically bring about greater gender equality and can affect women negatively (McIlwaine and Datta 2003, 370; Rathgeber 1990, 494), and there is the danger that greater involvement in the cash economy results in a devaluation of domestic work since it does not generate money (Lockwood 2009, 513).

Other research in the Andes indicates that *mestizas* working in public are frequently confronted with strong negative judgments and may be harassed by men (Bourque and Warren 1981; Weismantel 2001, 47). Jane Henrici reports that in Pisac, located about two hours by bus from Ollantaytambo, female market vendors have been accused of prostitution and witchcraft (2007, 87–90). When asking female craft vendors about how being a woman affected their work, I sometimes mentioned these findings; this almost always caused surprised looks, and the women said that for them this was not the case. While they did mention challenges in their work, these did not include harassment for inappropriate work but rather the struggle to combine their work with childcare and other domestic work. Similarly, Ferguson's work in Costa Rica, Belize, and Honduras shows that women are still the ones held primarily responsible for housework and childcare, even though combining these with the irregular and often late shifts of tourism employment can be difficult (2010b; 2011). It is also important to consider that, while selling crafts to tourists is a new type of work, it shows many continuities with the past. Andean markets have long been dominated by female vendors, and, even though their roles can be contested, women's positions in the public realm have a long history (Babb 1989; Weismantel 2001). Thus, while contact with foreigners has certainly increased significantly, many types of tourism work constitute modifications of what has been done for generations.

In Ollantaytambo, women almost always spoke positively about their involvement in tourism, yet there is also evidence of some backlash and conflict. Regarding women selling in the market, Diego said, "Yes, sometimes they talk with other locals, other vendors, and so people say, 'Ah, you are talking to him; you are with him.' So they fight at home, and the husband goes to drink." On a few other occasions, I heard criticism about women working "out there." Based on her experience running a hotel, Kate commented, "I had women work with me and then have issues because they were earning a good wage, and that made the husband feel a little bit threatened [...] also because she now has the opportunity to separate." Kate also related a story that indicates how tourism work can be linked with an increase in domestic violence:

> I was working with a couple from a small community about an hour away. She started in September and worked until just after Christmas; during that time her husband also started doing things in the garden and some housekeeping.[...] They were really good workers. But in their community, there were meetings when they weren't there, and so a new road was built through part of their garden. And in the end – he didn't seem like the kind of guy who would beat up on his wife; I don't think he was violent before – but I think it had all to do with not being there, not being present in the community. It was horrible.

In this case, it seems that it is not tourism work per se that causes people to act more violently but rather its intersection with existing gender roles. In other parts of the Andes, researchers have found that an increase in women's earning power can intensify patriarchal patterns in the home (Wilson and Ypeij 2012, 9). In Chiapas, Mexico, two female Mayan potters were murdered, likely because the wealth and status they had gained through tourism was perceived as threatening by men (Colloredo-Mansfeld 1999, 34). Across Latin America, the processes of globalization and economic deregulation have challenged the more traditional masculine roles. As women gain autonomy, men see their work options and authority become more precarious, which can lead to greater domestic violence. However, there is also a new trend of men increasingly sharing tasks previously defined as women's work (Olavarría 2006, 35–7). This can be seen in Ollantaytambo as well. As outlined at the start of the chapter, Diego willingly stepped in and looked after their sons when his wife was busy with her work. I came to know several local men, all of them in their twenties and thirties, who regularly looked after their children. Similarly, in Costa Rica and Belize, women who earned their own income could sometimes claim greater rights in the family and successfully request more help from men in domestic tasks (Ferguson 2010b, 870–1). Andrea Cornwall has criticized how development campaigns aimed at girls and women frequently frame men as a hindrance or threat (2014, 133–4). While men may oppose and restrict women's participation in the marketplace, many also show active support.

Furthermore, contact with tourists can also effect changes (Kinnaird and Hall 1996). Jason, a young American tour guide and long-term resident of Ollantaytambo, commented on how he viewed the impact of tourism on women's roles in the community:

> It gives women a lot more work opportunities. Traditionally here the women would stay home and cook and look after the kids, you know, and the men would go out and work. But now, especially with the younger girls, they used to have kids or get married when they were sixteen, seventeen years old. Now that things are getting more developed here you can actually meet twenty-five-year-olds who don't have kids and are working on their career.

He added that new work opportunities also existed in other sectors but that tourism seemed to be the driving force. As other researchers have pointed out, women's control over productive resources or the opportunity to earn and control wage income is strongly related to their status; it is also important that their culture's gender norms support them in these roles (Friedl 1991; Lockwood 2009). While women in Ollantaytambo are benefitting from greater income, their new roles are still controversial. This has been documented in other tourist destinations. For example, Jennifer Rodes's film *Trekking on Tradition* shows how the inhabitants of a mountain village on Nepal's popular Annapurna trekking circuit respond to the more casual relationships among Westerners and the greater independence of women. Whereas older people tended to be highly critical of this behavior and feared for their cultural values, several young community members expressed hope that their culture would change accordingly (Rodes 1992). I found a similar mix of responses in Ollantaytambo. While some criticized the behavior of foreign women, María, for example, emphasized the positive impact that hosting volunteers has had: "It's good for my children. They meet people from all over the world; they hear how life is in other places.[…] My daughter sees how independent many foreign girls are, and she can learn."

Another point to consider is how the different types of tourism work have distinct implications for women and men. Diego and his brother Alberto had started out singing and posing for tourists as children; now they are running their own tour company. María had worked as a craft vendor before converting her home into a guesthouse. Both Naida and Adriana had worked in high-end hotels; Naida and her husband Víctor had also been guides on the Inca trail, a four-day trek to Machu Picchu, before opening their hostel. When speaking about their work, many people emphasized that they valued workplace flexibility and control in order to accommodate childcare and socializing. As described earlier, Ana's work at the hotel had conflicted with childcare, whereas in her own restaurant, she could control her work more and have her children nearby. One of the photovoice images from Diego, not included here, shows his wife working in the kitchen with her two sons playing next door; in another photo, we see one of their sons with a toy truck just outside the restaurant. Both Ana and Diego mentioned the advantages of this arrangement several times. Since women are considered the primary caregivers for children, the flexibility to combine work and childcare is crucial for them to succeed and be supported by family members in their work. In several of the small restaurants I frequented, children were often present playing in a corner or doing their homework at an empty table, and frequently the owners visited with friends during the quieter hours. This blending of public and private roles can also be seen in small guesthouses where rental and living spaces are close together and allow women to attend to home duties while also being available to guests. Likewise, the craft market and small stores are dominated by female vendors. Infants are carried on their mothers' backs while smaller children can be seen sleeping on a pile of blankets tucked in a quiet corner. Children often join their mothers to eat a meal, and older ones may step in to attend to customers. Camila, a woman in her forties, commented, "That's good about this work here: we can bring our kids. Often my daughter comes by after school and she helps a bit. Sometimes she brings her friends and they play." Craft production and socializing can also be integrated into this flexible work schedule. Many of the women can be seen knitting wool hats or sewing bags, and during quiet hours vendors move their stools together, chat, and share a meal (see figure 3.1).

One of the photovoice images Daniela contributed, not included here, shows two local men and a woman sitting on the bench in front of her store. Daniela chuckled and told me they were relatives who "really like that bench" and were "always visiting." In another photo, one can see her six-year-old daughter in front of the store. Daniela commented that the girl liked to spend time there and was learning English from tourists. There is concern that the need for flexible schedules generally means less secure and poorly paid work for women (Olavarría 2006, 35), but in Ollantaytambo women emphasized that their small businesses generally afforded them more security and autonomy than did working as an employee. The greatest complaints about tourism work I heard from both men and women focused on strict and often unfair treatment by employers. Both Naida and Adriana described being harassed by supervisors at a hotel and threatened with job loss if they did not work extra hours, and, consequently, both of them had sought independent work in tourism. A couple of women mentioned that sometimes their husbands helped them with craft production, which was confirmed by volunteers working with weaving cooperatives. It has been documented that men may take over craft production once it becomes profitable for women (Little 2008, 161; Swain 1993), but none of the women I spoke with mentioned this as a concern.

Less than a generation ago, the majority of people in Ollantaytambo lived by farming (Hubbard 1990). Many people mentioned that their current work in tourism paid better and was less strenuous than agriculture, which is similar to Ferguson's findings from Belize and Costa Rica (2010b, 869). Yet there are also parallels between these types of work. In the Andes, agricultural work happens on a somewhat flexible schedule; it is often communal and can be combined with childcare. Likewise, Andean produce markets have long been social spaces where female vendors can bring along their children (Babb 1989; Weismantel

3.1 Market vendors socializing, by Karoline Guelke

2001). However, it is only the small, family-run establishments that afford this flexibility, so unless gender roles shift more significantly, only these types of work are easily accessible for *mestizo* women.

Marketing Indigeneity

For indigenous people, there are other factors to consider. The visual is an important factor in tourism, both in the representation of a site and in tourism practice. Certain cultural objects as well as the bodies of local people become prominent actors in the staging of tourist displays, including advertising, museum displays, official cultural performances, and individuals posing for photos (e.g., Dann 1996; Desmond 1999). Many of the concerns of the *incanismo* movement, such as the preservation of Inca architecture and the promotion of indigenous culture, converge with tourism and have been turned into marketable commodities (Silverman 2002; van den Berghe and Flores Ochoa 2000). Facilitated by community organizations and NGOs, indigenous people have increased their access to the tourist market to sell their weavings, leading to a greater validation of indigenous markers. In 2005, UNESCO declared Taquile Island and its handwoven textiles, produced largely by women, a Heritage of Humanity, which has brought recognition from beyond the community (Ypeij and Zorn 2007, 125). Several local people and volunteers commented that women weavers from highland communities were getting more empowered and confident through their interactions with tourists and the income their work provided. In this discussion, I briefly consider the issue of displaying indigenous clothing for decorative purposes and then explore the intersections of gender and ethnicity in the context of women and children posing for photos.

Diego chose indigenous-style knitted hats as wall decoration for the new restaurant. While these examples, shown in figure 3.2, consist of more modern designs, similar displays of traditional indigenous clothing items can be found in numerous hotels and restaurants in the Cusco area. On the one hand, it can be argued that this display of hats makes visible and honors indigenous heritage. Diego has relatives in the nearby Quechua communities and is well aware that these types of clothes are worn on a daily basis. He told me that he had seen a similar

3.2 Wall decoration in Ana and Diego's restaurant, by Diego

arrangement in another hotel and thought it looked beautiful. On the other hand, the items are removed from their daily use and have become simply decorative, aesthetically pleasing objects to brighten up a wall. In several places, I saw clothing placed alongside antique Inca tools or framed photographs of Machu Picchu and other archaeological sites, thus visually aligning them with the past. As mentioned earlier, this fascination with the past is a common theme in tourist imaginaries and "a replay of the evolutionary model, where things/people farther away are deemed to represent a more distant (and purer or more primitive) past" (Salazar and Graburn 2014, 10). This notion has also been documented in postcards from the area. Mary Weismantel writes about "a fantasy of pre-modern life" that creates a greater distance between depicted local people and the usually White viewers (2001, 180). The display of indigenous people's clothing constitutes a similar distancing and can obscure the fact that people in indigenous communities are active participants of the present.

Another way in which indigeneity is displayed is by people posing for photos. This activity is particularly common in Cusco, where, on many days, I encountered at least fifty women and children ready to pose, usually accompanied by a lamb or baby alpaca. Again, we find an alignment of present-day indigenous people with the past. The combination of women, children, and animals is also worth noting; these displays seem to position women as exotic, nurturing, and close to nature. Andean women are often seen as more indigenous than their male counterparts (Canessa 2005, 131; de la Cadena 1995), and in the context of tourism, they can become the "signifiers of traditional culture, the indigenous, and the 'Other'" (Babb 2012, 38). Analyzing the staging of people and animals in Hawaii, Jane Desmond writes that bodies are presented as natural, innocent, and authentic; people and places become dehistoricized, and "nostalgia replaces history" (1999, 254). Although in the Andes the sexual attributes of local women are not as emphasized, the themes of nostalgia, timeless tradition, and closeness to nature are clearly present.

In Ollantaytambo, one can also see indigenous women and children posing for photos, especially in the old part of town and near the entrance to the archaeological site. However, there are some important differences. Unlike in Cusco or Pisac, I never saw them posing with animals. The majority of women were from nearby highland communities, and they were posing in the dress still commonly worn there. On some days, one could also see groups of four or five women who would wait for a tour group to walk by and then sing a high-pitched Quechua song for tourists. These encounters were clearly mediated by the tour guides, who would explain briefly where the women were from and what type of song they would sing. After the short performance, the tourists usually took pictures and paid a few *soles*. People in Ollantaytambo showed a range of opinions about these performances. Rosa said that she found it sad that children were working, adding that they should rather attend school than come down into the valley to beg. Regarding the photo of indigenous girls (figure 3.3), Alberto commented that this work allowed people to buy goods they could not grow in their communities and that it was easier work than agricultural labor.

However, other local people disapproved. One local hotel manager commented that people were charging far too much. Her neighbor weighed in saying she was working hard in her hotel every day, whereas "they just stand there and ask for money." Resentful sentiments also came through in the comments of Adriana's father who told me that several NGOs had been focusing their work on the nearby Quechua communities and that the municipal government had recently paid for development projects there. "People up there make good money," he said, "but they still come down here and play poor."

Most tourists, on the other hand, expressed fascination with indigenous people, though their interactions were sometimes ambiguous. Lisa, the British volunteer, commented on her experience:

> At the plaza, I hadn't seen the girls in their traditional dress; it was our first week, and I was like "I love them; I really want a picture." And I always ask people, "Can I take your picture," because I know if I was walking around and people started taking photos I would at least want to have a say, so I always ask. And they were like "Yeah, yeah, of course," and then they are like "a *sol*, please." And of course I gave them a *sol*. I didn't agree to that, but then I felt bad, and I kind of realized then what the relationship was between us.

While Lisa had not expected to pay the young women, she complied. Dean MacCannell writes that the "touristic ideal of the 'primitive' is that of a magical resource that can be used without actually possessing or

3.3 Indigenous girls interacting with tour group, by Alberto

diminishing it," which sidelines the issues of economic disparity (1992, 28–9). As will be discussed in the next chapter, avoiding "paying too much" is a prevalent concern for many travelers. While it is easy to judge the tourists' attraction to these cultural markers as superficial, we should also reflect on the parallels with anthropology. Pierre van den Berghe even suggests that our discipline is "the ultimate form of ethnic tourism, the endless quest for self-understanding through the exotic Other" (1994, 32).

In some cases, tourists do not seem satisfied with what they are seeing and attempt to stage their own ideas of authenticity. Anja, the young German resident, described the following scene. Two girls dressed in handwoven clothing were posing for tourists, but another girl, clearly a friend of theirs and dressed in the store-bought clothes now common in the village, attempted to join the photo. The tourists, however, motioned her to get out of the way so that they could take a photo of the indigenous-looking girls only. As Michael Hill observed in Cusco, most tourists do not take photos of *mestizo* children in school uniforms but focus on those in handwoven clothing, since the children in traditional dress meet their preconceived ideas of the authentic Andean child (2008, 256). Likewise, one of Alberto's photos shows two girls, friends who were playing together while also posing for photos, and he commented that tourists just wanted photos of the girl in Quechua clothing (see figure 3.4).

Similar situations have been documented elsewhere. In the American Southwest, tourists have rearranged the hair and clothing of Native People before photographing them (MacCannell 1992, 28), while in Ecuador, they asked locals to pose in specific ways (Meisch 1995, 453). This behavior is also well illustrated in the film *Framing the Other*, in which we see a Dutch tourist taking photos of the Mursi people in Ethiopia, famous for the large lip plates worn by women. The visitor commands locals to smile and loudly calls out a woman as fake for not fully inserting her lip plate (Kok and Timmers 2012). Tourist brochures and websites are full of images of indigenous people, and travelers may want to reproduce those in a hermeneutic cycle (Albers and James 1988; Urry 1990). Elsewhere I have examined this effect in the context of tourists' photos posted online (Guelke 2014/2015). There are also continuities with historical photographs in which members of the middle and upper classes staged themselves in association with indigenous people while simultaneously reproducing messages of their own racial superiority

3.4 Two girls, one in typical Quechua dress, by Alberto

and modernity (Poole 1997). While encounters with indigenous people may briefly meet tourists' yearning for a past they never experienced, what do these encounters mean for local people in the present?

On the one hand, the interest of tourists provides some validation of indigenous markers, which is a reversal of the discrimination that people in handwoven clothing often experience, be it when a bus passes them by at a designated stop or when being served last and in a rude manner in the town's produce market. Florence Babb states that "what has been a social liability, being female and indigenous, can serve in some cases as a form of cultural capital" (2012, 38). On the other hand, the validation is literally quite superficial. Like "the Inca," the Peruvian from the coast posing in Inca costume, these women and children are staging themselves for the tourist gaze. Whereas "the Inca" knew multiple languages and could adjust his strategies based on the type of customers, however, the tourism expertise of indigenous people is generally more limited, especially given that the women involved in these encounters are often monolingual Quechua speakers. My conversations with them were almost always mediated by children, who are mostly fluent in Spanish and commonly function as translators with tourists as well. While this mediation can be very effective, interactions are still limited, and through their greater linguistic expertise, *mestizo* intermediaries often control the business of marketing indigeneity. Another point to consider is that most markers of indigeneity are easily staged for tourists. On several occasions in Cusco, locals there commented to me that the women posing for photos were "not real" and that their dress showed several nonlocal elements. *Mestizas* or *cholas* can easily dress up for tourists, who are unlikely to know the difference. I will return to issues of cultural appropriation and the power of intermediary agents when discussing New Age tourism in chapter 6.*

* Sometimes the gaze was reversed (Maoz 2006), and tourists became the objects to be photographed. One morning I saw a very tall and dark-skinned tourist of African descent surrounded by local people; they were laughing, posing next to him, and taking photos. When I later asked a local shopkeeper about this, he chuckled and said, "Yes, I *had* to take a photo to show my wife! So dark – did you see him?" Speaking about his work at the chocolate store, Jonathan commented that as a tall, blond foreigner with long hair, he felt "part of the spectacle." His partner Lisa described how, the previous day, ten Latin American tourists had taken photos posing next to her: "It feels a bit like being a character in Disneyland, you know?" Despite these occasional reversals, it is clear that, in tourist encounters, local people are usually the ones experiencing objectification.

Some researchers have voiced concerns about commodification and the potential loss of cultural meaning (e.g., Greenwood 1989). In the Andes, the colorful fringes along indigenous women's hats served to signal their social status, such as being married or widowed. Jane Henrici reports that in many indigenous communities around Pisac, people do not know the designs' traditional meanings anymore, as girls are increasingly wearing multiple colors in order to appeal to tourists (2007, 93). In Ollantaytambo, at least for now, women and children appear to pose in their traditional dress; the only additions I observed were a few plastic flowers pinned to the hat, which can be seen in figure 3.3.

It is also interesting to consider this work in terms of its intersection with ethnicity. Men from indigenous communities have been moving to urban areas to work at much higher rates than women, leading to strongly gendered rates of modernization (de la Cadena 1995, 343). Women's access to tourism work through posing for photos may help counteract this trend. However, as feminist anthropologists have emphasized, gender norms also have to support women's productive work and decision making if their status is to improve (Friedl 1991; Lockwood 2009). In other Andean communities, women's work selling produce or crafts is not valued as much as men's labor (de la Cadena 1995, 341–2), and in Cusco, posing for tourists is generally not considered a respectable occupation and can lead to family or community conflict (Babb 2012, 47). While I did not conduct research in the highland communities themselves, several comments by locals in town certainly reflected disapproval. However, Alberto also mentioned that community members worked on a rotation system, so that people took turns coming down into Ollantaytambo. Several people confirmed this point. This also explains why I rarely saw the same women and girls several days in a row. The fact that this system is in place indicates that there cannot be widespread opposition to women engaging in this work; instead, the communities have taken steps to manage participation and distribute the benefits more evenly.

Gendered Tourist Experiences

Tourism and anthropology are both rooted in imperialism, and the narrative of the intrepid, White, male traveler continues to shape both tourists' and ethnographers' experiences today (Dubois 1995; Elsrud 2006).

Western travelers often act according to their own gender roles or may actively challenge the norms from both home and host cultures (Myers and Hannam 2008). In the Victorian era, when travel and exploration were dominated by men, women risked their respectability when traveling alone (Elsrud 2006, 183; Enloe 2000). As in many cultures, being a woman meant staying close to home. But rather than seeing this gendered domestic-public dichotomy as a fixed system (Lamphere 2009; Rosaldo 1974), following Butler (1990), we can ask how gender roles are actively performed and constituted in the context of tourism.

By definition, tourists navigate mostly in the public spheres of the places they visit, and thus, female travelers in the Andes are diverging to some degree from the local norms. The masculine adventure discourse, which is so frequently enacted in backpackers' stories of difficulty and bravery, also obscures the fact that female travelers are generally at greater risk (e.g., Frohlick 2010; Jordan and Aitchison 2008). Compared to other South American destinations, Ollantaytambo is relatively safe, and men typically do not make aggressive advances. However, travelers do face certain risks, and in my sample, clear differences emerged in how men and women perceived and responded to potential danger. (These differences are depicted in figure 3.5.)

Tom, a young American backpacker, volunteered at a hotel in Ollantaytambo before continuing his trip around South America and Europe. When I asked him about safety concerns, he responded:

> Aside from basic precautions, like not flashing money around, I don't feel I need to be … I keep very little on me when I walk around town. I feel comfortable walking around here, even at night. Helps to be male, I suppose. No one has accosted me, aside from trying to sell stuff. I feel safe enough.[...] Jessica, the other volunteer, kept saying she wouldn't go out walking by herself after dark. I don't know; it's hard for me to say. As a guy, I never felt that way. I've been perfectly comfortable walking out in the streets at night in my own town and around here. It just doesn't occur to me that someone … But I don't know if she is just overly cautious. Others I've met are cautious, but they do move outside as long as there are other people. She didn't even want to go out with others.

3.5 Couple walking at night, by Karoline Guelke

Regardless of gender, most tourists mentioned similar "basic precautions" such as not carrying a lot of money, expensive watches or jewelry and not staying out late at night. Even though he voiced some doubts about the way the female volunteer restricted her behavior, Tom did recognize that his gender afforded him greater safety. Jessica herself had this to say:

> I think here in Ollanta, compared to Cusco and Lima, I feel more comfortable going places by myself. I mean, I don't go out after dark alone when I'm traveling, but I feel this is a pretty safe place for female travelers. I mean, everyone seems pretty nice; I haven't had any weird, awkward encounters with guys or anything like that. It has actually been really refreshing to travel here.

Even though she feels more comfortable and safe in Ollantaytambo than elsewhere, Jessica still adheres to her strategy of not going out alone at night. This degree of self-surveillance and restriction was mentioned only by women. When I interviewed couples, one partner often stated that they did not stay out late, but male solo travelers like Tom voiced far fewer concerns. The dangers were definitely real. As I will describe in chapter 6, Stefanie, a German backpacker in her mid-thirties, was sexually coerced by a tour guide in nearby Bolivia. She describes her safety strategies as follows:

> My habits here are different, definitely, because I don't know the codes. I'm more afraid because in the guidebooks you often read "watch out with the men, the guides," and all that. For example, in Germany I hitchhike a lot, but not here. And here I greet everyone, the men, also the women. But with men it's an issue of safety; I want a sort of relation, you know? If they greet me back, it feels better. I feel safer then.

Luz Marina, a dance teacher from Lima, commented, "Men always see a woman traveling alone as an opportunity, no? And the differences stand out, sometimes racial or how we dress. That attracts them generally." Only women mentioned changing their dress and covering up more; several European and American women said they felt uncomfortable wearing shorts. Yet at any given time during a warm day, one can see female tourists in shorts and/or tank tops, an issue that also came up in the photovoice project. As addressed previously, local people generally spoke positively about tourists, and much of their criticism was directed at men and women equally. Dress, however, was an exception. Daniela, the hotel and store manager introduced earlier, contributed the image of two female tourists (figure 3.6).

She commented that wearing shorts was not respectful and that women should have at least their knees covered. Anja, the young German resident, was told at the school where she worked that she needed to avoid short shorts or skirts and very tight clothing if she wanted respect from local people. She pointed out the irony that skin-tight leggings and T-shirts had become the common attire of local women, so the main issue seems to be the amount of skin showing. While Anja mostly followed these recommendations, many female travelers do

3.6 Two female tourists outside Daniela's store, by Daniela

not show much awareness of different cultural norms, and those who do make changes appear to do so primarily to avoid unwanted male attention. In the cross-cultural context of travel, aspects of gender performance can get lost in translation, so that, for example, the acts of walking alone and wearing shorts may be considered transgressive.

The experience of travel is gendered in other ways as well. Jonathan, the British volunteer, commented on an experience he had at a local festival with his partner Lisa:

> Well, Lisa being blond, White, and pretty, a lot of guys here want to dance with her. And the guys would come up to me and then kind of like befriend and then like ask me. It's kind of strange. One time Lisa was like, "No, thank you." And the guy, he heard her response but then still looks to me for an answer. And then I'd say, "*No quiere bailar*," you know, "She doesn't want to dance."

> And there's almost this – it's like a second validation; you need the man to say if it was or wasn't correct.

Similarly, an Australian backpacker mentioned that when she started a conversation with locals, they often directed their responses to her boyfriend, even though her Spanish was much better than his. On the one hand, traveling with a male partner provides greater safety and freedom for women, but it also seems to restrict their interaction with locals to some degree.

Female solo travelers, however, often commented that they enjoyed the contact with local people. Yet while travel can be an empowering and emancipatory experience for women, the greater risk also results in higher levels of self-restriction, counteracting some of these effects. In many places, female travelers are criticized far more than males for transgressing social boundaries and for behaving in ways considered overly sexual (Frohlick 2010). A study in Italy found an increase in self-surveillance among female travelers, who restricted their movements and bodies in order to avoid harassment from local men (Jordan and Aitchison 2008). Daniela judged the women's clothing in the photo as inappropriate, and on several occasions, local women commented that I was a good person because I did not stay out late in the evenings. Frequently, they contrasted my behavior with that of female volunteers from North America or Europe who had spent many evenings at the bar and associated with local men. However, overall, the gendered criticism I heard was limited, and, as I will explore in chapter 6, in many cases, locals showed more concern that foreign women would be taken advantage of by local men. Interestingly, sometimes female travelers can also benefit from a certain gender ambiguity. Because female travelers look different from local women and travel alone, their gender performance may be perceived as more masculine than feminine. On several occasions, people in Ollantaytambo pointed to a tourist, usually a short-haired woman in long pants and hiking boots, and asked me to clarify the tourist's sex – was this a woman? On a visit to Cusco, I once passed a young boy who gazed at me wide-eyed and loudly asked his mother whether I was a man or a woman. I believe it was my tall stature combined with bulky clothing, as well as the fact that I was walking alone, that in his eyes made my gender performance ambiguous.

The self-restriction reported by most female travelers is also shared by LGBTQ+ tourists. While Cusco and other South American cities have specific LGBTQ+ destinations, this is not the case in rural areas. A German tourist in her forties told me that she was interested in meeting local women but found it difficult. At home, she knew where to go, she said, but in Peru she was not sure and felt afraid of being judged or even attacked. An Australian backpacker had traveled with a female partner for a while but had made an effort to hide their relationship for fear of negative attention. That had also been the approach of Ade and David, the Indonesian couple. When they stayed at her guesthouse, María referred to them as "the two friends," seemingly unaware of their relationship. Ade commented, "Yeah, don't ask, don't tell. It's just easier that way." While the liminal stage of travel can afford greater freedom in terms of sexual and other behaviors, women and LGBTQ+ travelers in particular need to discern carefully which contexts are safe and which may put them at greater risk.

As we have seen, gender plays out in multiple different ways in the context of tourism. It has been widely shown that paid labor is generally valued more than unpaid labor, which means that domestic work and childcare, largely performed by women, do not receive the same recognition as men's work outside the home (e.g., Lockwood 2009, 513; Rathgeber 1990; Waring 1988). This also applies to Andean regions (de la Cadena 1995). In many parts of the world, development programs have aimed at creating opportunities for women to enter the marketplace and receive monetary income in order to improve their status (Chambers 2010; Kinnaird and Hall 1996). The work opportunities in tourism have similar potential and can arguably contribute to women's empowerment. They may also allow for more personal agency than development programs imposed from the outside (Aslanbeigui, Oakes, and Uddin 2010, 191). In the context of gender and development, empowerment has been defined as a situation in which people previously limited are able to make a wider range of choices (Kabeer 1999; Mosedale 2005). The fact that so many women in Ollantaytambo are running small tourism businesses implies that they were able to make this choice and are not being restricted in significant ways. In turn, the flexibility of small businesses facilitates a greater range of choices than the more structured employment available in larger businesses. Although their public position has sometimes been challenged, women have dominated as market vendors in the Andes for centuries, so their work in craft sales constitutes an extension of a traditional occupation rather than a radically new development. Furthermore, we need to recognize that people regularly enact different identities, and Andean market women have long performed different roles to adjust to specific situations and customers (Babb 1989, 25; Henrici 2007). For example, they may modify their dress and language in order to alternatively emphasize their indigenous characteristics or their belonging to an urban setting (Weismantel 2001, 115). Thus, as they now interact increasingly with international customers, market vendors may add new roles to their repertoire, but this does not necessarily imply a loss of cultural or gender identity.

Discussing patriarchal gender relations, Naila Kabeer points out that "power and dominance can operate through consent and complicity as well as through coercion and conflict" (1999, 441), and patterns of normalized inequality can be difficult to identify for the people involved. Aspects of *machismo* and *marianismo* mean that family and community support for women is greatest where their new work in tourism allows for a combination with existing domestic roles, especially childcare. Since women generally allocate a greater share of their income to the household than men (Olavarría 2006, 35), supporting them in their work outside the home benefits other family members as well. Working in small, family-run restaurants, guesthouses, and craft stalls has allowed people to carry over a certain flexibility and sociability characteristic of agricultural work; yet, as we see an increase in larger tourism businesses, local women may not be able to participate and benefit as much. Sarah Mosedale defines empowerment as a collective process rather than an expansion of individual women's choices only (2005, 252). Similarly, development initiatives have been criticized for imposing the Western notion of an autonomous individual, which is opposed to the more collective Andean view (Apffel-Marglin and Sanchez 2002). The way in which indigenous groups near Ollantaytambo are communally managing women's participation in tourism seems to be a successful example of such a collective approach. In the valley, however, tourism development also brings about an increase in inequality and conflict, which I will turn to in the next chapter.

4

Negotiating Material Inequalities

I met Markus and Karina, a Swiss couple in their early thirties, over breakfast at a small guesthouse. While we were sharing fried eggs, bread, and jam, Markus told me that the two of them were backpacking around the world for about one year. Karina commented on the stress of traveling and her frustrations with feeling overcharged: "It's exhausting to be ripped off; I mean, it's not about the money, but after a while you start feeling really bad. Do they think we are stupid and don't notice?"

Yet, the couple also recounted an experience that reflects a different attitude. The previous day the two had shared a taxi with an older American couple to visit the Inca site Moray about a half hour's drive away from Ollantaytambo. The American man had haggled relentlessly with the driver, even after Karina and Markus felt they had reached a fair price. Markus commented, "For that money no taxi driver in Switzerland would even start his engine. We ended up giving the driver a bit more money on the side. I mean, he was wearing … he looked like he could really use some money."

Karina and Markus stayed for another day and then continued their trip to Bolivia. On Facebook, I could follow their journey and see their stunning photos from other parts of South America, Australia, and South East Asia. After returning to Switzerland, they posted reflections about their trip, saying that they had realized that most of the people they met when traveling would never be able to leave their own countries and that they now appreciated their own opportunities and privilege much more.

Tourism often involves unequal relationships, and in Peru, the most obvious material inequalities exist between local people and foreign visitors and residents. It is important to acknowledge that "tourism ... both reinforces and is embedded in postcolonial relationships" (Hall and Tucker 2004, 2). Mary Louise Pratt speaks of "contact zones," defined as "social spaces where disparate cultures meet, clash, and grapple with each other, often in highly asymmetrical relations of domination and subordination – like colonialism, slavery, or their aftermaths" (1992, 4). One of these social spaces is tourism, which consistently brings together people of different cultural and social backgrounds. Despite their country's economic growth, over 20 percent of Peruvians were still struggling with severe poverty in 2015, most of them in the Andean highlands (Oxfam 2017). In Ollantaytambo, some inequalities between foreigners and locals are obvious: the European tourist with the large camera around her neck ordering a meal in an upscale restaurant, while a farmer in well-worn clothes and rubber sandals passes by on his way back from the fields. Other inequalities are more hidden. My focus in this chapter is on the ways in which tourists and locals perceive and negotiate some of the main inequalities; in chapter 6, I will return to this issue and examine conflicts between tourists and long-term foreign residents, as well as conflicts among community members themselves. My findings indicate that tourism perpetuates and exacerbates certain forms of inequality, yet these are often mediated by practices of solidarity and cooperation. Understanding what kinds of situations and conditions facilitate the latter can help guide tourism development in more equitable ways.

CONCEPTUALIZING INEQUALITIES

The inequalities I focus on in this chapter refer to the general gap in material wealth between locals and visitors and, contingent on this, the power differential people experience in everyday situations.

While acknowledging the broader systemic inequalities, I also draw on a Foucauldian understanding of power as fluid and continuously enacted and challenged (Foucault 1978, 1980). One of the main ways power relations are produced and exercised is through discourse. The ways in which people consistently speak about a certain topic develops into commonly accepted views of how things actually are; thus, discourse can be seen as a way of constructing truth and reality (Foucault 1972). In the context of tourism, the discourse about specific destinations in developing countries is generally dominated by Westerners who often have the power to tell other people's stories (Bruner 1991, 241).

In addition, Pierre Bourdieu's three different types of capital can help us conceptualize the different intersecting inequalities. Economic capital refers to access to material resources, while social capital describes a person's group memberships and institutionalized networks. Cultural capital consists of skills and knowledge gained through education but also includes aspects like language and dress. The three kinds of capital are strongly correlated and affect a person's social mobility (Bourdieu 1984). These categories were conceived of in the context of the stratified but culturally very homogeneous French society of the 1960s and 70s, so when applied to cross-cultural encounters, the picture becomes more complex. Fifty dollars translates into very different buying power in rural Peru versus the United States, yet economic capital can still be compared relatively easily. The dimensions of social and cultural capital, however, are very much culturally constructed; what types of social connections and skills are available and valued vary strongly between cultural contexts. Therefore, we cannot simply conclude that one person has more than another but need to consider the different context as well, a point I will return to below and in further chapters.

Following Pierre Bourdieu's categories, we can say that the majority of tourists visiting Ollantaytambo have more economic, social, and cultural capital than residents do. In terms of economic capital, some of the traveling vendors may constitute an exception, though usually only temporarily. Travelers who fund their trips by selling self-made jewelry and other goods usually get by on a very low budget. This is also noted by local people. Valentina once commented to me that she felt sorry for the "poor travelers," some of whom she had seen sleeping under tarps in the plaza. Although these travelers may certainly hit hard times financially, they generally have social and cultural capital that allows them to negotiate these potential difficulties. For example, a young Chilean backpacker I spoke with had a master's degree in philosophy. As we talked, he switched from Spanish to fluent English and later threw in a few phrases in German. His pants had large holes in them, his hair was long and tangled, and he looked like he had not washed in days. While sitting on the dusty steps of the plaza offering his handmade jewelry, he called out to passersby in different languages and drew them in with intriguing stories about the meaning and healing power of the stones. He had been traveling through South America for almost a year and was considering heading to Asia or Europe next. "Or I could go home and get a job," he laughed, "but not yet." While he may have been temporarily low on money and in some ways looked similar to the Andean farmer passing by, in terms of social and cultural capital, he clearly had much larger resources at his disposal. In addition, being White afforded him other privileges as well.

Material poverty is a hard reality for many people in Ollantaytambo and the Sacred Valley. As I came to know local people better, I heard many of them referring to their past in terms of struggle. Diego, who recently started a tour agency with his brother Alberto, described their childhood as follows:

> When we were growing up, there was a lot of poverty. We were six brothers, and often there was not enough to eat. We had herb tea for breakfast and a piece of bread; that was it. Lunch was a soup, sometimes with a little meat, but usually without, and we only had a real main course about once a month.[...] So we kids had to help with making money. My brother and I would go to the ruins and play flutes and sing, dressed in our ponchos. We started doing that when we were seven, eight years old; we basically grew up in the ruins. We've been working in tourism ever since.

For many people, the challenges are ongoing. Johnny, the street vendor, lives in the nearby town of Calca, and on most days spends an hour on the bus to come to Ollantaytambo to sell his art. After I had come to know him over the course of a few weeks, he invited me to the birthday party of his girlfriend Catalina's son. Before we joined the celebrations, we walked through Calca's market to pick up a few bottles of soft drinks. Johnny pointed to a shoe-shine boy no more than eight years old and said, "That's what I used to do when I was little. I would come back from school and then work until evening." After the celebration, Johnny and Catalina took me to their home. We entered a muddy courtyard, flanked by a rough concrete wall and a long two-story building with about twelve separate rooms; one outhouse and water faucet were shared by everyone. The second story was only accessible through wooden ladders, and as Catalina climbed up in her high heels in front of me, she turned around and said apologetically, "This is where I live – unfortunately." I followed her into the one room that she shared with Johnny and her son; it had one bed, a wooden table with four chairs, clothes stored on a shelf and in various plastic bags, and a couple of toy trucks on the floor. After a generous meal of fried duck, potatoes, and pasta, Johnny and Catalina stopped to talk with an old woman in their courtyard, who was weaving on a back-strap loom. As I bent down to admire the intricate design of the wide belt she was working on, I caught a glimpse of her small room with nothing but a mattress on the floor and a few large bags with personal belongings hanging from the ceiling. The three were conversing in Quechua, and later Johnny told me that a middleman would likely pay her the equivalent of eight dollars for the piece, which takes at least three days to complete. Tourists in the Cusco area will witness certain expressions of poverty, but most of them are unlikely to get a glimpse of, or imagine, what daily living conditions are like for many of the people with whom they interact or whose crafts they purchase.

TOURIST DISCOURSE

Cusco's center has changed drastically in recent years as more expensive stores, restaurants, and hotels catering to the more upscale tourist market have displaced smaller businesses. In any other of Cusco's neighborhoods, however, social inequalities are immediately obvious to a visitor. Whether by train, bus, or taxi, the trip from Cusco to the Sacred Valley leads through the fast-expanding marginalized neighborhoods on the hillsides above the city. Many residents have recently migrated from rural areas; construction is booming while infrastructure is clearly lacking. The material poverty is most visible in the unfinished houses and large amounts of garbage in the streets. Many precariously constructed dwellings cling to the hillsides, connected by meandering and often muddy footpaths, though in recent years new concrete staircases have replaced some of these. Many people sell food and other goods from small, makeshift roadside stalls.

On my frequent bus trips through this area, I repeatedly witnessed reactions from tourists as they encountered these sights. Usually, they had worried expressions as they looked out the window, expressing sentiments such as "Wow, this looks pretty grim" or "Geez, sure wouldn't wanna live here." (See figure 4.1 for a depiction of the tourist gaze.) I also noticed that many tourists quickly turned their attention away, pulling out cell phones and guidebooks or engaging in conversations with each other. The tourist gaze becomes unsettled in places like this. These are not the images foreigners recognize from tourist brochures but a sudden and unsettling confrontation with material poverty more familiar from "World Vision" brochures. Unlike in official slum tourism (see, e.g., Frenzel and Koens 2012), where a guide directs, and thereby justifies, the tourist gaze, here there is no clear script of where to look. Interestingly, the sight that usually elicited the greatest response was a large pile of garbage frequented by dogs digging around for something edible; generally, several tourists got out their cameras or cell phones to quickly take a photo. Conscious or not, focusing on animals instead of humans can serve as a strategy to distract oneself from the impact of witnessing poverty, a point I will return to below.

Aside from pity and a certain fascination, fear was another common reaction from visitors who passed through this location. For a couple of months, traffic in the area often slowed to a crawl due to a major construction project to widen the road. As we were inching along one morning, a street vendor with a baby on her back approached the bus to sell food through the window. A middle-aged man in our bus loudly advised his three travel companions to beware: "Hold on to your cameras! Let's hope we don't get stuck *here!*"

As the bus continued its way out of Cusco and descended into the picturesque Sacred Valley, cameras and cell phones came out again,

4.1 Tourist looking out from bus, by Karoline Guelke

this time to document the farmhouses surrounded by fields and eucalyptus trees, framed by mountains rising steeply on either side. As on the outskirts of Cusco, there are people showing signs of material poverty, but tourists' comments here almost invariably focus on the landscape: "Wow, beautiful" or "Look at those mountains – amazing." Even though rural areas suffer disproportionately from poverty, this may appear less obvious to a foreigner because it is embedded in bucolic landscapes and impressive mountain scenery; poverty seems more "natural" here and even becomes romanticized.

The same process may be in effect in Ollantaytambo, which, despite undergoing rapid tourism development, still preserves a more rural and quiet atmosphere that contrasts sharply with the urban environment of Cusco.

One of my standard questions for tourists was about the differences they noticed between themselves and local people, usually followed by a more specific question about their experiences encountering poverty, if any. Some comments indicated that tourists were not fully perceiving or acknowledging the material inequalities. An older Dutch man commented that "they seem to be poor, but everybody has their smart phone as well," indicating doubt that the poverty witnessed was real. On different occasions, tourists showed an inability to imagine different life circumstances and poverty. In some cases, they seemed to focus solely on how the poverty of locals affected them rather than considering what the disparity may have meant for the people themselves. On one of my bus trips between the valley and Cusco, I spoke with a middle-aged German couple traveling around Peru by themselves. The woman recounted several positive experiences and referred to Peru as "such a beautiful country." When asked about her experiences with poverty, she responded by saying that she felt "worn out" by seeing it but that it was not as bad as the begging they had experienced in the Caribbean. "One can't give to everyone, so right now we don't really give anything. Yesterday a woman approached me when I was sitting on a bench in the plaza and asked for money, but I said no." The issue of poverty was translated into begging, which is the aspect that affected her. She also recounted a time she and her husband had visited the San Pedro market in Cusco. Apart from a few souvenir stalls, the market caters largely to local customers, selling anything from household goods to clothing and ritual and medicinal items. The material poverty of both the neighborhood and many of the vendors, some of whom sit right in the street with a handful of items laid out on a blanket, is clearly evident. Yet the German couple's comments reflect a different experience: "We went to the San Pedro market; that was pretty good, but the streets below, now that was something – very funny all the different things they sell. And there was this woman with a little child on her back, and the child was chewing on a chicken foot! In Europe they'd call child protection services!" It is interesting that, even though I had asked specifically about her experiences with poverty, she emphasized the "funny" aspects of the experience and the chicken foot, certainly an unfamiliar food to a German but common locally and no reason for concern. Focusing on a minor issue such as that can be a way of avoiding the larger problems of inequality and suffering.

Most tourists recognize material inequalities but may still justify or distance themselves from these in different ways. In an analysis of tourist postcards in Cusco, one of the common themes identified in the depiction of rural children is "poor but happy" (Sinervo and Hill 2011), a view that also emerged strongly when I asked tourists about their perceptions of local people in Ollantaytambo and the Cusco region more broadly. Lisa, the young British volunteer introduced earlier, spoke enthusiastically about local people, saying that "they are so nice, so happy, even though they don't have much." Similarly, the Dutch man quoted previously told me,

> They seem quite poor, but they are making something of their lives. When you come from a Western country, then you realize this is a really different life. But people seem to be rather happy with their lives. I don't know why, but they seem rather cheerful and friendly. I'm really impressed by their attitude and way of living.

While these comments reflect admiration for local people, they also indicate an emotional distancing. If people are happy, then the poverty witnessed may not be as severe, so tourists need not worry about being implicated and can simply learn something about finding happiness. Michael Hill interprets this view as "an emotional accommodation and rationalization for class and race hegemony" (2008, 265). Other comments combined the "poor, but happy" theme with calls for intervention. An American man, who was traveling around Peru and Bolivia with his own guide, commented, "People look content to me. It's definitely a much simpler lifestyle, but people seem to enjoy their lives and their families.[...] They seem to be getting by okay but definitely need help in some areas." Despite their happiness, he considered people unable to manage their own lives and in need of outside assistance, presumably from the Western world. Renato Rosaldo writes that "imperialist nostalgia occurs alongside a peculiar sense of mission, the white man's burden" (1989, 108). The phrase "they need some help" was used by a number of other tourists I spoke with, though many volunteers also reflected more critically on this and questioned some of the local aid projects. Barbara Heron points out that the concern for helping the subjugated is not just a remnant of colonialism but a process whereby Westerners continuously perform their identities as good, moral people (2007, 9).

In addition to being ignored, dismissed, or pitied, people experiencing poverty are also romanticized, which achieves a similar emotional distancing—the "poor, but happy" attitude. The "exoticized tourism imaginary" includes both the "conceptualization of the exotic subject in terms of primitiveness and lack of civilization" as well as the "noble savage, living in harmony with nature" (Theodossopoulos 2014, 60), as we have already seen in the case of indigenous people posing for photos. Poverty can make for captivating images. One morning I was standing in the doorway of María's guesthouse when a few meters away I saw an older man lying in the street. His back was propped up against the stone wall, and there was a large empty bottle beside him. A moment later two tourists came around the corner; they stopped for a moment, the woman lifted her camera to take a photo of the man, and then the two quickly continued on their way. All this took no more than a few seconds. I was struck by this brief encounter and decided to paint it later (see figure 4.2).

How had the tourists perceived this man? Leaning against the ancient Inca wall, dressed in torn old-fashioned looking clothing, he in many ways embodied the image of the poor, drunken Andean man. The scene was reminiscent of photos taken by Martín Chambi, the Peruvian photographer who became famous through his portraits of indigenous people in the first part of the twentieth century and whose images are still found on every postcard rack today. For most tourists familiar with these ubiquitous images, the scene would have had a familiar and timeless quality, again, reproducing the common theme of local people as living in the past. In the context of tourism, poverty and struggle become exoticized and subsumed in a romanticizing gaze. For example, in the Caribbean, American tourists find the local beach shacks quaint and appealing while in their hometowns they tend to stay away from poorer areas (Pruitt and LaFont 1995). Susan Sontag has stated that "to photograph people is to violate them, by

4.2 Tourists photographing local man, by Karoline Guelke

seeing them as they never see themselves, by having knowledge of them they can never have; it turns people into objects that can be symbolically possessed" (1977, 14). Does my drawing run the same risk? I, too, depicted the man in a vulnerable position, which again points to the parallels between the tourist and the ethnographic gaze. However, as discussed in the introductory chapter, I believe that the power differential is somewhat minimized through the artistic method: by foregrounding the interpretations of the anthropologist-artist and the incompleteness of the scene, the image can reflect ethnographic information without claiming the same authority or power as a photograph.

How do tourists experience their position of privilege? Michael Hill speaks about tourists experiencing "First World Guilt" when encountering people in poverty (2008, 261). An Australian woman, traveling around the world with her boyfriend, commented that sometimes she hid her camera not so much because of safety concerns but because openly displaying her wealth made her feel uncomfortable: "I've got this big camera and a tripod. It's so flashy, like, look at all my money, you know? So I prefer to carry it in my backpack." A similar sentiment was echoed by a middle-aged Dutch woman, who said she did not wear much of her jewelry, not because she felt particularly threatened but because she did not want "to show off." I, too, had acted similarly. During an earlier period of fieldwork in 2002, I had fallen and torn my favorite pair of pants. Nevertheless, I continued wearing them and somehow thought that it might make me look more similar to local people, who often wore well-worn or damaged clothing. However, the local woman I was staying with at the time, a former schoolteacher in her sixties, admonished me harshly and told me that it was inappropriate for me to be seen like this. It was, of course, rather naive of me to think that a tear in my pants would somehow bring me closer to locals and magically erase the much more fundamental material inequalities between us.

As mentioned, when they were confronted with views of poverty, tourists on the bus often directed their attention to animals, which can be seen as another strategy of emotional distancing. Animals were also a common theme in conversations with tourists, volunteers, and expats, many of whom voiced concern about specific dogs and cats and, in some cases, adopted them. Walking around the village, one frequently encounters dogs, but in most cases these are not street dogs; they have owners who allow them to roam free during the day. Peggy, an older American resident, had taken in an abandoned dog and walked him regularly on a leash, something commented on as unusual and amusing by several locals, one of whom referred to the dog as "her boyfriend." For the photovoice project, Daniela took two photos of tourists interacting with dogs (not included here) and commented that "sometimes tourists seem more interested in the dogs than in the people." One evening over dinner, I was talking with María and her husband Ernesto. María mentioned that foreign volunteers staying with them had adopted two of their kittens and taken them abroad, and a friend of hers had given one of her dogs to an English guest. "These animals see more of the world than we do," joked Ernesto.

As described in the opening anecdote of this chapter, Karina said that it was "exhausting to be ripped off." I frequently heard travelers complain about people trying to sell to them, and matters of price, getting good deals, and avoiding "getting ripped off" were central themes in conversations. Jessica, the American volunteer spending three weeks working at a local hotel, told me how disagreeable she found this experience:

> I didn't expect that people constantly want to sell me things. I've been here for two weeks now, and the same people are still trying to sell me stuff. Oh my God, it's so frustrating.[...] I didn't expect that to be all over town. It's a little jarring, but I guess it's no different from how it is in Cusco or the other tourist cities. So I'm trying to get used to it again: right, whenever you are traveling, when you are a White person in a foreign country, that's sort of the experience.

Jessica's comment shows frustration but also indicates some awareness of her position of privilege. Many tourists, however, are simply fixated on getting the best deal. Kelly, a Canadian in her sixties, regularly spent the winters traveling on her own. We had met at María's guesthouse, and I had offered to take her on a hike to Pumamarca, a site of pre-Inca and Inca origins located about two hours by foot up the valley. Before heading out, we stopped in the produce market to buy some snacks, and I watched as Kelly walked back and forth between stands to inquire about the prices of bananas and avocados. The main produce vendors did not budge, but after several minutes Kelly found a vendor outside, a woman in torn clothing selling some fruit spread out on a cloth in front of her, who agreed to sell for less. Five minutes of bargaining had saved Kelly the equivalent of about 50 cents, and she had benefitted from a person who was clearly materially worse off than she was. The strong focus on paying as little as possible was common. Jacquie, an American in her early forties, had been running a hostel with her local partner for a couple of years. She described experiences at her guesthouse as follows:

> What pains me to see with tourists is when they don't have a sense of how to be generous. I wonder how much of that is affecting the local people in terms of becoming more greedy. So many of the tourists I've seen, they have plenty of money. Maybe in their country they are middle class, not really rich, but they can travel, and then you have a local person asking for a tip for a photo, and they don't give anything. Also, at breakfast – when I was talking to guests, finding out what people needed that day – almost 90 percent of the conversations focused on money: where they went and how much they paid for this or that, just money-money-money. It's such a prevalent part of their conversations; it's just amazing. So I would tell my guests, "Think of the universe, or God or whatever, giving you an opportunity to be generous!"

There are other aspects to consider as well. Several tourists commented on feeling stupid for being overcharged; as quoted above, Karina said that "after a while you start feeling really bad. Do they think we are stupid and don't notice?" Lisa, the British volunteer introduced in the previous chapter, commented, "It makes you warm less to people, you know, because you think, 'Is he gonna rip me off?' Sometimes I wait for the vendor to say the price to another Peruvian, so that I know. I really try not to be like a dumbo, you know?" On the other hand, bargaining over prices is also customary in Peru and functions as a form of social interaction. As addressed earlier, many tourists are seeking some connection with local people, and haggling over prices can be one way to engage in conversation. Clare Sammells has interpreted Western tourists' bargaining in Bolivia as an attempt to display their knowledge of local customs. In hopes of getting closer to indigenous culture, tourists emphasize their own lack of money by performing "a camaraderie of poverty" (2014, 133). Last, in line with the idea of indigenous cultures and practices as located in the past, "bargaining may be a fun, exotic game that conjures up a more 'primitive' and less alienating exchange system far removed from the impersonal and unyielding price tags of commodities in Western stores" (Hill 2008, 264).

It is also evident that many local people do charge tourists more, which is commonly referred to as "*gringo* prices." During one of my first bus rides between Ollantaytambo and Urubamba, the largest town in the valley, the young bus driver told me a price three times higher than the regular fare. When friends of mine were visiting, we took a taxi to a nearby Inca site, and after we arrived, the driver insisted on a price that was significantly higher than what we had agreed on earlier. Nevertheless, the time and effort many tourists invest in getting good deals seem out of proportion to the economic and other forms of capital most of them have available. As Barbara Heron suggests in her analysis of development workers in Africa, the strong discursive emphasis on feeling targeted and treated unfairly can serve to distract from the visitors' power position based on Whiteness (2007, 98).

Traveling on the cheap can also serve purposes other than saving money. Giulia, a German-Italian backpacker of about forty, told me that she always tried to connect with local and traveling street vendors since they were knowledgeable about cheap places to eat and stay. Hazel Tucker makes an interesting point about this, observing that traveling independently and outside of organized tours increases the opportunities for chance encounters and experiences, such as meeting people on a local bus. Thus, traveling on a shoestring is not always merely a strategy for saving money or a sign of stinginess but can be used to increase opportunities for serendipity (2003). Giulia's comments, as well as the experiences of other travelers who highlighted chance experiences based on "roughing it," support this point. In a study conducted in Cusco, tourists also tended to judge their travel experience as more authentic if it involved the difficulties and challenges typical of low-budget travel (Hill 2008, 265). This finding is in line with Nelson Graburn's model of travel as a ritual of inversion: for middle-class tourists, temporarily giving up their familiar comforts becomes an important aspect of the trip (1989). Stories of challenge also form a significant part of cultural capital for tourists, and the greater the challenges the more admiration one can obtain from one's peers.

LOCAL VIEWS OF INEQUALITY

As Darya Maoz reminds us, the tourist gaze is not a one-way street; local people gaze back and form their own views of tourists (2006). Jason, a blue-eyed and cheerful American in his late twenties, had lived in Ollantaytambo for about five years and worked as an adventure guide. I had seen him often with his large backpack when leaving for or returning from multiday hikes, and we met later when I stayed in the guesthouse owned by one of his friends. One evening, I visited him in the old Inca house he rented. After he had cleared away empty beer bottles from the table, we settled down for the interview. Here are his comments on locals' perceptions of foreigners:

> We have to remember that everyone who makes it here comes from a socioeconomic level to afford the ticket, getting at least two weeks off work; they can travel. So it's very interesting, sometimes my local friends will ask me, "Are all buildings in the US really tall? Does everyone have a car?" Well, many people have cars, yes, but we also have people standing in line to get free food at the soup kitchen.[...] It's interesting talking to friends of mine from here who have gone to the States, and they come back and they are like, "Dude, you were right. You're not all tall and rich and blond-haired, and people sleep in your streets, too!"

While some Peruvians have the opportunity to see that poverty also exists in the United States and other countries, the fact remains that, in material terms, most Westerners who visit Peru are far better off than local people. It is also worth noting that definitions of poverty vary cross-culturally. Annelies Zoomers writes that, in the Andes, poverty has widely been defined as a lack of land and of social connections, and in the southern Peruvian Andes, this view began to shift only in 1990 when NGOs introduced the concept of material poverty (2008, 976–8).

Encounters with tourists certainly demonstrate a significant wealth gap to local people. Travelers frequently compare how many countries they have visited, and high numbers can translate into cultural capital among peers, but local Peruvians also often asked me which countries I had visited. My list includes twenty-two countries, and this answer usually elicited surprise and astonishment and made me feel acutely aware of my privilege and relative wealth. I was also painfully reminded of O'Rourke's classic film *Cannibal Tours*, which shows encounters between Western tourists and locals in New Guinea (1988). Near the beginning of the film, a German tourist clad in safari gear recounts the long list of countries he has visited. Speaking with an air of arrogance and self-satisfaction, he immediately establishes himself as the unlikable "ugly tourist." Since I did not want to be seen like this, I began to reduce my list of countries when people asked me. While this naive strategy helped ease my own discomfort temporarily, obviously it did nothing to alter the underlying inequalities that separated us. In one instance, I was called out on this unequal opportunity explicitly. I was walking around the community of Patacancha in the highlands above Ollantaytambo, looking after the young daughter of a friend who was spending the day working there. As we stopped to pet a cat, an older man stepped out of his small adobe house and shouted a greeting at us. He then asked me where I was from and which countries I had visited. I mentioned about six, and he slowly repeated the names. Then he said emphatically, "You are lucky. I would like to travel, too, but I'm poor!" In *Cannibal Tours* the European and American tourists cruising up the Sepik River express a number of different theories about what they perceive as the backwardness of local New Guineans, whether viewing them as happy primitives in a state of nature or as representing a lower evolutionary level. In contrast, local people clearly recognize the decisive factor that separates them from the visitors: economic resources. In one scene an elderly woman says angrily, "White men got money; you got all the money" while in another an old man states, "If I had money, I could travel on that boat, too."

SOLIDARITY, CONSIDERATION, AND REFLEXIVITY

Not all tourists are out to get the best deal at all times; many expressed concern for local people and tried to be generous in different ways. As Jane, a British nurse in her sixties, commented, "Sometimes we just paid a bit more. One time a woman in a small village charged us 15 soles for three coffees. That was too much, but whatever; we paid it. It's still cheap for us. I mean, who would you rather be ripped off by: a poor Peruvian or a bloody big company?" A similar understanding was expressed by Jonathan, the British volunteer:

> You know, I don't mind it so much. They are like entrepreneurs, trying to make some money. It's not like Europe where if you don't work you get support; if you don't work here, you starve to death. So they're kind of doing what they can.... I try and have that in my mind when I'm walking along and people are pestering me. They are trying and making a living, you know.

Giulia, the German-Italian backpacker, partially funded her travels by selling handmade jewelry that was very similar to the work of local Peruvians. When I asked her about this, she said, "Yeah, I was thinking about selling some of it here in Peru. But why compete with the locals? I'm not rich, but these guys need the money more than I do." Similar to Jane and Jonathan, she indicates a greater awareness of the underlying inequalities, and her comment also reveals that she has backup funds or other means of making money. Hill describes how tourists in Cusco are motivated to purchase goods based on the "First World guilt" they feel; shopping thus becomes framed as a way of helping local people (2008, 262). He points out the irony of this, yet local people told me that this is exactly what they wish tourists would do. One of the most common complaints about visitors I heard was that they are "just passing through" and "they take a look but don't buy anything." In a neoliberal context, the solution to poverty is considered to lie in increased commerce.

Sometimes tourists give support in more direct ways. Jacquie, the long-time American resident, has encouraged this; she told me, "Tourists have asked me if locals are offended if you offer them money, and I say *never* is a Peruvian offended if you offer them something; that's what their whole culture is based on. You go to a highland community; they have five potatoes left, and they'll give you all of them." Jacquie refers to Andean customs of reciprocity and sees tourists as potentially fitting into this system. However, simply giving out money can clearly be problematic. Kate, the British hotel manager, described the following situation:

> There's one example of this guy who literally had a fistful of dollars and was giving it out to the children. And I'm like, 'Hey, what are you doing?' And he said, 'These people don't have any money; I'm giving them money.' And so, you know, it was really hard to explain to him that this wasn't a good thing, that this was really setting them up as beggars.

Tour guide Jason also commented that giving out money perpetuates the stereotype that every foreigner is wealthy and can afford to give. However, Kate's example also illustrates the helplessness some visitors may feel when confronted with poverty. Others choose to give money with strings attached. Debbie, an American tourist in her fifties, recounted how on a recent three-day trek, she and her group had passed through small highland communities and met an old woman who was walking barefoot. Their guides Diego and Alberto had told them that she lived alone and did not even have money for shoes. Debbie decided to give the woman 10 soles (about 3 US dollars) but asked the guide to tell the woman that she had to use the money to buy shoes. Even though she did not know anything about the person, Debbie assumed she knew best what was needed and did not grant the elderly woman the agency to make her own decision about the money given.

While some tourists seem unaware of their own privilege, many also expressed awareness and gratitude. In their Facebook post, Markus and Karina talked about how, after their travels, they had come to appreciate their own opportunities and privilege much more. Michael Hill has also identified the theme of tourists feeling "blessed and fortunate" about the wealth and comfort in their own lives; however, he adds that they rarely recognize how they and their home countries might be implicated in these structural inequalities (2008, 264). Stefanie, the German backpacker mentioned earlier, showed some awareness of this:

> I travel very slowly, and I visit many places that are not so touristy, but still I ask myself, "I'm a tourist, too, no?" When I travel to less touristy places, then tourism grows there as well. And sometimes it seems like the second ... like the Spanish, we are like the Spanish, no? First they came and now the tourists. It brings a lot of changes and many problems.

Her comment indicates a broader historical and postcolonial reflection; she does not simply individualize her experiences. Like Hill (2008), however, I found this consciousness to be the exception; while many tourists expressed concern for local people and reflected on their own privilege, I seldom encountered an awareness of larger global and historical connections.

Some have argued that the dichotomy between tourists and hosts may be overemphasized and that many international travelers establish deeper connections and even long-term friendships with local people (Wilson and Ateljevic 2008). We can certainly see some examples of that in Ollantaytambo. Naida said several times that her positive experiences with guests outweighed the bad. She told me that when she and her husband first opened their hostel, they had no internet presence. Several of their first guests convinced them of the need for this service and helped them get set up on Tripadvisor and Booking.com. Later, an American told them they needed better photos online; he then took pictures of their rooms and lobby and posted them. Naida also mentioned that they started with no computer knowledge but often guests helped them out, in one case spending hours setting up a Wi-Fi connection for them. On several occasions, the brothers Diego and Alberto emphasized the positive side of tourism. While they had grown up with the constraints of material poverty, their children are now attending school and are well taken care of.

Local people also expressed concerns over tourists being overcharged. One craft vendor told me, "I don't like how some people take advantage of tourists. I know that not all tourists have lots of money, but many local people think they do." A young woman from a nearby community, the only woman among the otherwise male street vendors of paintings, said that she thought the entry to the Inca site was overly expensive for foreigners and that "they should be more lenient with students. Sometimes they forget their student cards, but they don't have much money." Daniela, the photovoice participant and hotel manager, often criticized local taxi drivers for overcharging tourists. Her concern for visitors was also evident in several photos shared with me (see figure 4.3).

She explained that photo:

> This is the oldest woman in town. She doesn't know how old she is, but she must be in her eighties. She still sells food on the corner there in front of our hotel. She does have family, but she wants to contribute; she still wants to work.[...] The tourists in the back there – the traffic congestion was so bad, they had to get out and walk down to the train station to catch the train. Some of them were elderly, too.

For months I had seen the old woman sit on the street corner, always with a large bundle next to her, but I had not realized that she was selling food. No sign indicated this; she was simply selling to locals who knew her. In her bundle, she had two or three pots with a home-cooked meal, as well as plates and cutlery; she served the meal, her customers would eat it sitting on the steps nearby and then return the plate to her. This is another example of a small-scale business that can be run with great flexibility and low investment, as I have discussed in the previous chapter. Also noteworthy is the juxtaposition of elderly locals and visitors in Daniela's comment; while the local woman is still working, elderly tourists enjoy their leisure time. Despite the difference in wealth and privilege, Daniela shows empathy and concern for the visitors.

While in their praise of tourism many local people focused on economic capital, other aspects played a role as well. The brothers Diego and Alberto emphasized that they enjoyed learning about other cultures by talking with tourists from all over the world: "This way we have educated ourselves; it's a benefit, an interchange of culture. We

4.3 Old woman and tour group, by Daniela

4.4 Cards and gifts displayed in María's guesthouse, by Karoline Guelke

cannot travel because we do not have enough money, but this way we can inform ourselves and learn." I heard similar sentiments from other people working in tourism, who saw their interactions with foreigners as the only way to learn about other places. On the wall in her kitchen, María displayed photos and thank-you cards she had received from volunteers who had stayed with her (see figure 4.4). She was proud and protective of the special cabinet she had reserved for gifts, including mugs, plates, packages of tea, and a bottle of whiskey. This display of social capital immediately showed off her international connections to anyone who entered her kitchen. While this may have given her some status in the community (as well as exposed her to envy, addressed later), these social connections were often limited. When one evening María mentioned that a couple of Canadian volunteers had invited her to Vancouver, I spontaneously said, "That's great! You can visit them and then come stay with me in Victoria."

While I was momentarily oblivious to the restrictions she faced, she simply looked at me and said, "Yes, in another life."

Despite the underlying inequalities, there are also moments when locals and visitors can come together and simply enjoy each other's company. In Ollantaytambo, most accommodation still consists of small locally owned guesthouses where a family home has been reconfigured to rent a few rooms. Family life is not kept separate from guests; children run about and converse with guests, family friends and neighbors drop by for a chat while the owner is preparing breakfast, and guests are sometimes invited to family celebrations. I spent a whole evening of New Year's celebrations with Naida and Víctor and a Brazilian couple who were guests at their hostel. We shared a meal, played games, and toasted to the New Year; afterwards, everyone agreed that it had been a great evening. On another occasion, María told me about her sister's birthday; María's husband had been away that evening, and with her sister being unmarried, they just had female guests. Scrolling through the photos on her cell phone she chuckled as she described the evening to me:

> We were just sitting around that evening, you know, drinking a bit, talking ... and then Lucho, the kid who lives with us and runs bicycle tours for tourists, he brought in these seven big guys from England! I had my *comadres** here, in their traditional skirts and all, and at first they looked really upset. But then we started dancing, and it was so much fun! And my *comadres* started dancing, with their skirts swirling, and with these big guys jumping around ... oh, I laughed so hard, my stomach hurt! We had so much fun!

Small informal family businesses like María's and Naida's guesthouses foster closer contacts between hosts and guests. In the context of rural Turkey, Hazel Tucker found that through the close encounters afforded by small guesthouse accommodation, local people can control some of the tourists' behaviors and engage with them on their own terms (2003). These settings allow people to come together, enjoy

---

* The term *comadre,* literally co-mother, refers to a woman connected through a godparent relationship (the male version is *copadre*). Sometimes these terms are also used for close friends.

each other's company, and simply have fun – despite their underlying differences and inequalities. While these momentary encounters and shared experiences usually do not translate into long-term connections, they often constitute "meaningful encounters" (Wilson and Ateljevic 2008) for all parties involved.

Tourism development in the Cusco region has been shown to increase material inequalities, as benefits from tourism are distributed in unequal ways (Steel 2013). In this chapter, I outlined how, in their daily encounters, locals and foreigners negotiate some of these inequalities in the "contact zone" of tourism. While there are clear structural differences with regards to the economic, cultural, and social capital of tourists and locals, the specific ways in which people experience and manage these vary strongly. Tourists may romanticize local people or construct them as exploitative, thereby discursively achieving distance from the poverty witnessed. As Clare Sammells suggests based on her work in Bolivia, in the context of tourism, poverty can simply become subsumed under the umbrella of cultural difference (2014). Just as Mary Louise Pratt emphasizes in her concept of the colonial "contact zone," in tourism as well, encounters tend to happen in a framework of inequality (1992). On the other hand, visitors also express empathy, consideration, and reflexivity. Local Peruvians are well aware of the different inequalities at play, and, despite the emotional labor required in tourism, they generally voice appreciation of visitors. The inequalities described in this chapter are often handled with "mutual complicity" (MacCannell 1992) and solidarity, yet there are breaking points where envy and conflict are expressed more openly, which I will turn to in the next chapter.

5

Conflict, Resistance, and Witchcraft

About halfway through my fieldwork, I began to suffer from acute insomnia. Often I could not go to sleep until three or four o'clock in the morning, and many nights I only drifted off briefly when the sun was coming up. Within a short while, I started feeling very anxious and depressive, especially since I could not find a cause. I was staying at the same guesthouse as before, and, if anything, felt more comfortable in the community now that I had come to know people. Occasionally I had a good night's sleep but overall I grew increasingly desperate as time went on and sleeping medication proved mostly ineffective.

When I mentioned my insomnia to local people, they almost always gave one of two explanations: that I had insulted *Pachamama*, or Mother Earth, in some way, possibly by unknowingly stepping or urinating in a sacred area, or that someone had given me the "evil eye." The second explanation was also the result of a traditional technique of diagnosis, which Valentina, Adriana's mother, performed. One evening, I went to her house, and, as I stood quietly in her kitchen, she slowly passed a raw egg up and down my body. She then cracked it into a glass of water and pointed to the white swirl that was visible. "Look," she said with a frown, "this shows that someone has given you the evil eye [*mal de ojo*]." When I inquired further, she suggested that someone in the community was probably envious of me and had caused my insomnia and that since I had "the soul of a child," I was particularly susceptible to such curses. This unsettling diagnosis was the first time I had heard of the "evil eye" in Ollantaytambo, but soon I found out that these beliefs were not uncommon. After about ten weeks my sleep finally returned to normal.

Beliefs and practices of the "evil eye" are found across southern Europe, the Middle East, and Latin America. The idea is that someone's gaze, motivated by envy or resentment, can cause psychological and physical harm to another person. Tourism development contributes to growing economic inequalities among community members, and those who succeed often become targets of hostility. We can distinguish "open resistance," which involves aggression and possibly organized protest, from more indirect "veiled resistance," which includes strategies like obstruction, gossip, insults, and humor (Boissevain 1995, 14–15). In this chapter, I examine how open and veiled resistance are expressed among community members as well as toward foreign long-term residents, who often bring expertise that enables them to be more successful in tourism businesses. This resistance can take the form of supernatural beliefs, practices such as the "evil eye," and rituals performed with the intent to cause harm. While I never found a clear cause for my insomnia, the explanations given by local people indicated that some of my behavior could have attracted envy and aggression. Last, I return to locals' relationships with tourists and the ways in which locals manage and challenge their visitors. While it seems that tensions are on the rise, some of the expressions of conflict, including witchcraft, can also be interpreted as attempts to restore balance to a community that is experiencing rapidly increasing economic inequalities.

CONFLICTS BETWEEN COMMUNITY MEMBERS

As a consequence of rapidly growing tourism and other forms of economic development, inequalities are growing between community members. American traveler Hubbard, who stayed in Ollantaytambo in the late 1980s, describes the typical houses as follows (1990, 66–7):

> I was somewhat shocked when I first saw the house interiors. They share it with their animals at night. A few of the village

> homes have wooden floors and are neat and tidy. But most homes in Ollanta ... have packed earthen floors, black from centuries of chicken droppings and baby urine and spilled kerosene and the blood of butchered rabbits brought inside to gut and dress. Hunks of horrible-looking meat hang from the rafters, where vast shoals of flies make their home. Most village houses have one bare light bulb hanging from a roughly hewn rafter.... Only a rickety table with uneven, handmade legs might grace the home as furniture. Everyone seems to sleep in one giant bed under mounds of dirty blankets.

His description has a dark and critical tone and clearly reflects his own uneasiness with the unfamiliar living conditions. Despite his bias, he nevertheless provides some vivid detail about the look of the village roughly three decades ago. American architect Graham Hannegan also commented on the changes in construction. Having lived in Ollantaytambo for over ten years, he witnessed many of the developments firsthand and described how the universal thatch roofs changed to tile and more recently to a number of new materials, reflecting the growing material inequalities in town. He also referred to the irony that the locally used Spanish term for concrete is *material noble*, the "noble material," indicating the view of concrete as a superior and high-status material.

Economic change has been as uneven as it has been swift. Even though the town has a rule that allows only two-story adobe constructions, larger buildings made of concrete have been built as well, often next to houses that look similar to what Hubbard described. (See figure 5.1 for a bird's-eye view of the old town.) One morning I was chatting with María and commented on the dilapidated house across the street from her guesthouse. "You know, they don't even have electricity over there," she told me. "The kids are often hungry; sometimes I give them bread and jam. It's very sad. Some people here are getting very rich from tourism, but others don't get anything." American resident Kate commented that she was seeing many people making money quickly while others remained poor, resulting in greater disparities within the community. These observations are backed up by national statistics showing that, while poverty rates have declined over the past decade, the wealth gap has stayed roughly the same

5.1 View of old part of town, by Karoline Guelke

or even increased (Oxfam 2017). In a small community like Ollantaytambo, the sudden and unequal inflow of money has fostered rifts between family and community members. Street vendor Johnny described one example to me:

> There's a family in Cusco, friends of ours, they were very poor and lived by selling postcards to tourists. Then they met this old guy, a foreigner, maybe American. He drank a lot and had money, and he gave them lots. They started building themselves a big house with concrete walls and all. Damn, he must have given them tons of money.[...] Now they don't talk to us anymore.

When I asked why, he added they became selfish and did not want to share their wealth. Interestingly, he mentions the house built of concrete as a key factor in how the family set itself apart. Foreign residents made similar observations. Jacquie commented that "ambition has really increased, in terms of money… and therefore people are having more disharmony in their families because they are squabbling more over things. That's what I've been told." Graham added that local men often spent their increased earnings on alcohol, leading to greater conflict within families. With regards to household organization, studies have shown that resources are not always shared evenly within families (Benería 2003, 35; Moore 1988, 55–6). Cultural power structures within the household can result in varying levels of intra-household cooperation and conflict (Costa and Silva 2010; Quisumbing 2010; Rai 2011b).

Within and between families, disharmony and conflict emerge in different ways. Many people commented on envy and gossip as major concerns. Based on her study in Göreme, a small town in Turkey, Hazel Tucker highlights the role of gossip as an informal yet powerful means of social control (2003, 108). María's comment echoes this: "When you do well, people talk. They don't like it. They might say you've dug up some Inca gold or stole or something to get rich." A local restaurant owner also described being criticized for his success:

> There's a lot of jealousy here. My childhood was not very good, and I wanted something better for myself and my children. But people here see that I now have a restaurant and a hotel, and to my face they are all nice and say, "Congratulations!" But behind my back I know they talk: "Oh, he is way too ambitious. Who does he think he is?"

In addition to this form of veiled resistance, sometimes the aggression and opposition become a lot more open. When Naida and her husband moved into town a few years prior and opened their guesthouse, neighbors were hostile; some yelled at them and even threw rotten eggs at their door. Although this open resistance has subsided, other conflicts are ongoing, which Naida explained by their neighbors' envy of their success. She also mentioned that another hostel owner down the street routinely turned their sign around and tried to entice visitors to stay at his place instead.

Despite this competition, goods and services offered are often remarkably similar, which was identified as a problem by the municipality (Sariego López and Moreno Melgarejo 2011). Craft stores and stalls tend to carry the same range of products, and restaurants often have very similar, if not identical, menus, a point criticized by many tourists. Foreigners often interpret this practice as indicative of a lack of knowledge and entrepreneurship. Tom, an American backpacker, had complained to me a number of times about what he considered the inefficient running of the hostel where he was volunteering. "Man, if I had more time, I could get this place working so much better," he said. However, we need to remember that definitions of efficiency vary between cultures and that the focus on maximizing material profit is linked to a capitalist mindset. It has been shown that in many peasant communities, people work hard with the goal of remaining at the same socioeconomic level, disregarding new ideas and innovations (Bailey 1971, 20–3; Foster 1965). In a cultural climate that values equality, the strategy of copying can allow people to keep up with changes without outcompeting others and further increasing inequality (Tucker 2003, 111). It was clear that in Ollantaytambo many people did not aim for maximum monetary gain. For example, to my surprise, a jewelry maker told me that during the high season he works less and on some days does not open his store at all. He explained that since sales are good at that time, there was no need to work as hard, and he could spend more time with his family. Thus, while we see clear examples of competition, this seems to be balanced by an emphasis on egalitarian values and social connections.

These values are also expressed in the many examples of cooperation. Most community members know each other and are connected by kinship, work, or cooperation in community events and festivities. These bonds foster a support network that extends into tourism work as well. Many times, I saw how hotel owners send guests to a friend's establishment when their rooms were full or they did not want to take any more guests. Craft vendors guarded each other's stalls or sent customers over to a friend. One vendor told me, "We are like family; we support each other." Another added, "When someone is sick, we collect money to help." When the artist Johnny had problems with his girlfriend and did not draw for a while, friends gave him some of their pictures to sell. The traditional Andean principle of *ayni*, or reciprocity, remains strong, a point I will return to below.

LOCALS AND FOREIGN RESIDENTS

Some tourists and volunteers decide to extend their stay, and a few eventually become long-term residents. Even those without extensive funds find that the far lower costs of living afford them a comfortable lifestyle. A retired German living in a community near Ollantaytambo commented, "Isn't life great in the valley? We live like kings here!" In neighboring Ecuador, the inflow of foreign residents has been shown to increase land prices and further impoverish local farmers (Gascón 2016), and this is also a problem in parts of the Sacred Valley. Ollantaytambo, on the other hand, does not allow the sale of houses and land to foreigners, so my discussion here focuses on residents who are renting properties and running businesses.* Often the specific behaviors of foreigners cause resentment from local people, as American resident Jason explained:

> I hate to say it, but some volunteers do contribute to the hostility towards foreigners. They come in for one or two months; they don't really stay long enough, but they start to act like a local in town, like they're entitled. Most people here are happy to give discounts to a volunteer, because they know they've been here for a long time and do help. But some just start demanding that, like get a free cookie, get a free coffee.... Whereas a lot of locals, the real locals, they're like, "No, you've been here for a month; you are not part of this community.[...] You do not know us; you do not know how this place works."

My own observations confirmed that long-term residents do not necessarily develop a greater understanding of or closer relationships with local people. One example of this was Peggy, an older American, who had lived in Ollantaytambo for about eight months but spoke little Spanish and had limited contact with local people. I frequently ran into her in the street and the small shops, and one evening I accompanied her to watch dances performed in honor of the local patron saint.

---

* I heard that some foreigners find ways around this rule by buying property in the name of local residents; conversely, some locals have sold land regardless, only to have foreign buyers find out afterwards that they did not have any legal rights.

The plaza was crowded with people. When a young boy pushed his bike past us and inadvertently hit the back of her legs, Peggy yelled loudly, "Jesus Christ!" – highly inappropriate at a Catholic celebration. On another occasion, I mentioned some of my difficulties with a local family to her but stated that I still considered them friends; Peggy almost shouted back at me, "Peruvians aren't friends!" I argued that I certainly felt I had formed friendships, but she insisted that most local people were just out to take advantage of foreigners. Comments by locals indicated that Peggy was considered unfriendly. "She's been here for a long time, but she doesn't talk to us," a store vendor told me. "She is rude."

Apart from specific behaviors causing bad feelings, the major frictions between local and foreign residents arise because of business competition. Almost all foreign residents I met had found work in the tourism sector, and based on their cultural background and education, they often brought greater expertise than did local Peruvians to this work. Jacquie described how she had first helped out at a local guesthouse and later opened her own hostel with her local partner:

> We were really successful really fast because I knew more what tourists wanted. The Peruvian people just don't see those kinds of details. They don't think of these things because they've never lived like this. You know, every successful tourist business here in town has had a *gringo* involved somehow.[...] I'm here not as a business woman, I mean, I need a livelihood, but it just appeared. I didn't even have to think.

Jacquie's North American background alone provided the cultural capital to help her succeed in the tourism business. Another example is Jason, the adventure guide. "I'm doing well," he said. "Tourists like me because I'm the perfect bridge, the American guy that knows this place, knows the culture. Many Americans just prefer an American guide." While cultural capital varies strongly cross-culturally, the processes of globalization and economic growth put foreigners and locals in direct competition, and in the arena of tourism, foreigners often win out (Steel 2013). This is clearly noted and criticized by local community members, including María:

> There is a lot of competition; foreigners just have more skills. They speak English; they know how to deal with internet bookings and advertising and all that; we usually don't. And foreigners help each other. They told you the best Wi-Fi is at the American hotel down there, didn't they? And that the Italian place makes the best pizza. Those are all foreign businesses, right? No one says, "Valentina's cakes are the best; stay with this local family." They just help each other but don't recommend people from here.

The business success of foreigners in the tourism sector is based on both their cultural capital and their social capital, since the strong support network they have among each other allows them to advance further.

The tensions and resistance are also expressed in more open and violent ways. American resident Ben had lived in Ollantaytambo for nearly ten years. Almost three weeks after opening his guesthouse, he still had not hung a sign outside and explained to me that he feared people would just deface it out of envy. His concern about animosities was not unfounded. Late one night at his guesthouse, I heard some commotion and yelling downstairs. In the morning, Ben looked tired and worried; he apologized for the noise and explained what had happened. For several years, he had participated in one of the local dance groups that perform in the four-day festival honoring the local patron saint. Several of his "friends" from the group had appeared at his door at night, highly inebriated, and challenged him to fight. Angered by the success of his newly opened guesthouse, they had yelled and told him he had no right to be there. It is significant that this open aggression is directed against a foreigner who, by any standards, is well integrated into the community. Dancing for the patron saint is considered an honor and an important service to the community; inviting a foreigner to do so indicates a significant level of acceptance and inclusion. The alcohol involved clearly contributed to the open aggression, but it is noteworthy that the accusations were voiced in this specific way, clearly positioning Ben as the foreigner who had no right to live and run a business in town.

Several other foreign residents had experiences of facing animosity and conflict, including insults and public attacks. Although the bartender Joaquin was Peruvian, his coastal origins and more European look made him a *gringo* in the eyes of many locals, which resulted in similar conflicts. He described how local men would come to his bar and sometimes yell at him, complaining that he was taking work away from them:

> You know, one of them was a friend of mine; we had worked together. Then he came by the bar one night; he was drunk already and claimed that I owed him money. He yelled at me, "You garbage; you piece of shit! This is not your land; this is our town!"

It seems that, despite high levels of integration and cooperation between locals and foreigners, the underlying inequalities remain and can lead to sudden outbursts of resentment and violence. Conflicts with foreigners may also become violent because the more indirect strategies of social control, such as gossip, are less effective with them (Tucker 2003, 109).

It is worth noting that there were different perspectives on the conflicts in the bar. As Jacquie commented,

> It feels like recently we are getting more *gringos*, including people from Lima.... Now they are like screening; they are closing the doors and screening, letting some people in and not others. It's supposed to be according to how drunk they are, but it often turns out that it's all *gringos* in there. My partner and I've been there several nights, and he's the only local! So, we are starting to see more of that as the *gringos* move in and take over the businesses.

This view was confirmed by Naida, who told me that the managers of the bar just catered to *gringos* and discriminated against local people. Indeed, Joaquin mentioned that he priced the beer higher in order to discourage locals from coming. A similar situation has been reported from Cusco, where owners of tourist restaurants in the city center have refused service to Peruvians, likely because they tend to consume and tip less than tourists (Steel 2013, 243).

Another factor is the sale of drugs, particularly cocaine, which is produced in the nearby lowland areas. In its entertainment section, the Moon Handbook states, "Ollantaytambo has a small but lively nightlife scene, due mainly to a community of expats who live in town year-round. Be sure to patronize places that seem respectful of surrounding residents and do not encourage use of drugs, which is an increasing problem in town" (Wehner and del Gaudio 2011, 46). Ironically, the one establishment recommended is exactly the one I repeatedly heard criticized for drug sales. This very real concern over increasing drug use further fuels animosities toward foreigners.

Alongside this growing competition, however, there are also many examples of cooperation and solidarity between foreigners and community members. Kate, who first came to Ollantaytambo eleven years prior and now runs a successful hotel, described her approach as follows:

> It is really important for me to run my business in a sustainable way. So I do have employees come from outside, but I also have several local employees who get well above minimum wage; they are all on health insurance. So, a woman who is pregnant will get maternity leave. And also, I do classes with them, not myself, but a friend of mine is teaching English classes, and I sponsor those; I think that's important.

Jacquie, a friend of Kate's, confirmed this, emphasizing how by educating local people through her hotel business Kate has given back to the community. Foreign-owned restaurants and hotels all employ local people, and I heard several owners express concern about involving and properly compensating them. Despite these efforts, tensions nevertheless appear to be strong and likely on the rise alongside the increasing material inequalities in the community.

WITCHCRAFT AND THE "EVIL EYE"

Conflicts also play out in the arena of supernatural beliefs and practices. When I was suffering with insomnia, Valentina diagnosed the "evil eye," suggesting that someone had been envious and thus made me ill. These beliefs date back to colonial times when the Spanish transferred European notions of witchcraft to the Andes (Silverblatt 1983). I use the terms witchcraft and black magic to refer to supernatural beliefs and practices that are intended to cause harm. While traditional Andean beliefs focus on maintaining balance between people and their environment, Europeans superimposed dualistic distinctions between good and evil, including their views of witches as having a pact with the devil. Interestingly, the figure of the devil has taken on more ambiguous connotations in the Andes and sometimes even functions like a patron saint (Silverblatt 1983, 421; Taussig 1980). In many parts of the Andes today, including the Cusco region, we find Andean beliefs and black magic existing side by side. Local newspapers include advertisements for "black shamans" whose services range from magic spells to bond a romantic partner to curses causing illness and even death. Despite playing an important role for local people, black magic remains largely hidden from tourists, and it was only after five months of fieldwork that people began speaking with me about this.

One afternoon, I was chatting with María on her porch. We were both concerned about Naida, a common friend, who was suffering from the return of a worrisome illness. For María, the cause was clear. "It's witchcraft," she said. "Why would she get it now? I think it's her husband's ex-wife; she is jealous." As Edward Evans-Pritchard states in his classic work on the Azande, witchcraft arises in the context of social relationships and as an explanation of misfortune (1937, 106). When I asked María if cases of witchcraft were common in the community, she nodded and recounted several instances when people had performed rituals, or asked someone else to, in order to harm someone, usually through illness or business failure. Another woman told me that she had repeatedly found burned candles and other paraphernalia on her property. There were disputes about land ownership in her family, and she suspected that a cousin was performing black magic in order to take land away from her. As mentioned earlier, female market vendors in the nearby town of Pisac were reportedly accused of witchcraft (Henrici 2007). The cases I witnessed in Ollantaytambo did not seem to target women specifically but largely reflected conflicts related to the growing material inequalities among community members, regardless of gender.

Foreigners can also become targets of these practices. After I returned from a few days away in Cusco, American expat Ben told me that two of his neighbors had seen a man walking around the block

5.2 Man performing ritual to cast a curse, by Karoline Guelke

three times in order to put a curse on him and his guesthouse. The man had been reciting something while carrying a candle and ringing a small bell. He was a former employee who felt he had been paid unfairly, though Ben claimed that they had had a good relationship and that the man had been well compensated. I was struck by the image and decided to paint it later (figure 5.2). As described in chapter 1, this art is an example of my depicting what I did not see but heard about (Causey 2017, 147).

The cases of suspected witchcraft I heard about all occurred between people who had fairly close ties through kinship or friendship, or at least a positive work relationship. Hazel Tucker has observed the same in Turkey. She writes that beliefs in the "evil eye" can be considered another expression of the striving to maintain equality that is typical for peasant societies. Black magic is a more formalized way to push down people who stand out and do better, especially in economic terms (Tucker 2003, 105). This belief also fits with the notion of "limited good," the idea that resources are finite so that one person's gain means another's loss (Foster 1965). Similarly, on Taquile Island in Lake Titicaca, those who expressed desire for greater wealth were seen as a threat to community harmony (Mitchell and Eagles 2001, 23), and in other parts of the Andes, wealthier community members have been accused of deals with the devil (Zoomers 2008, 977). The rifts of growing material inequality may be felt especially strongly between people who have, in the past, shared a close and more equitable relationship.

I, too, had felt that my connections with people were generally friendly and positive. Though diagnosed as suffering because of the "evil eye," I never found out if someone was, in fact, taking measures to do me harm, but the fact that a local woman diagnosed this condition indicated that something about my behavior was considered inappropriate and thus made me vulnerable. Valentina's comment about me having "the soul of a child" also raised interesting questions. On another occasion, when I expressed upset about a man yelling, María told me the same thing, with what seemed like a mix of criticism and affection. One evening over dinner preparations, Hilaria, a retired schoolteacher I was living with for a while, recounted a story based on an experience a friend of hers had had when visiting a hacienda several decades ago. As was often the case, the hacienda owner strongly guarded his daughters, but this particular case had been extreme. Her friend had told her how on a visit she had met all three daughters, grown women by that time, and found they were all dressed like young girls. The oldest one, then in her late twenties, had been sewing. Her friend assumed that it was for her baby, but it turned out she was making a dress for a doll. After finishing the story, she looked at me and said, "And that's how I see you, like this big girl. Forgive me, but to me you are this tall, beautiful girl, not really an adult." When I asked her to explain, she said something about me being innocent and pure but also naive. The image of the women dressed as girls stayed with me and made me reflect on the ways in which I may have appeared to some local people. The pen and ink drawing shown in figure 5.3 is based on this particular story but also ties in with Valentina's and María's comments about my childlike soul.

What are the implications of being seen as a child? How does being considered childlike compare to other anthropologists' experiences? Like a child, the ethnographer is learning about the cultural norms under study, which naturally involves awkwardness and social mistakes. The same has been documented for tourists and study-abroad students (Ascione 2020, 26). Local people also told me that they believed most foreigners had not encountered the same challenges they had. In addition, as an unmarried, childless woman of forty, I did not conform to local gender roles. One of the doctors in Cusco I consulted about my sleeping issues had a similar suggestion. After some basic tests, he asked me a few questions and then commented that the reason for my insomnia must be the fact that I was single and without children. His diagnosis differed from what locals in Ollantaytambo had suggested, but there was a common thread: they saw my health problem as related to my social role. I was not fulfilling my gender role in a culturally appropriate way, and therefore I was suffering. Other female anthropologists have addressed the issue of identity and roles. Reflecting on her research in Chiapas, Mexico, Christine Eber writes that her identity seemed to puzzle people, and a young boy described her as a "tall, pale, woman-like person who sleeps alone" (1995, 75). As in my case, other community members had compared her to a child, indicating that she was not seen as fully occupying the expected adult feminine role.

The second common view about my insomnia, having inadvertently insulted Mother Earth, reflects the notion that I had upset the balance with natural forces, and it was suggested that I perform an offering in order to rectify my transgression. Irene Silverblatt writes that in Andean communities, "notions of illness and well-being were intrinsically tied to a normative structure in which the maintenance of balance between social, natural, and supernatural forces was a predominant ideal. The explicit expression of this ideal is found in the term *ayni*, which means both balance and reciprocity" (1983, 418). While she describes the region at the time of European colonization, the

5.3 The anthropologist seen as a child, by Karoline Guelke

principles of balance and reciprocity have been documented in many contemporary Andean communities (e.g., Bolin 2006, 1998; Greenway 1998), and these principles were evident in Ollantaytambo as well. As described, traditions of support and cooperation continue strongly in the community, but this network is increasingly under threat by the growing material inequalities, something that has been reported from other communities in Bolivia and Peru as well (Zoomers 2008, 976). In Ollantaytambo, some of the aggression toward foreigners is clearly a defense and open resistance against outsiders' economic domination. But there are also attempts to integrate foreigners. The suggestion that I perform a ceremony to restore balance with Mother Earth is an example of this, and even the potential "evil eye" can be seen as a strategy to "bring me down" to the local level. Attributing my sleep issues to lack of compliance with the culturally acceptable role implies that there is the possibility of integration and connection. However, the situation with tourists who stay for only a few hours or days is different.

LOCALS AND TOURISTS

As we have seen, the relationships between tourists and local people are usually marked by significant material inequalities, yet both groups are often complicit in performing interactions in a way that sidelines these differences. Most tourist encounters I witnessed and heard of were characterized by a friendly tone. However, as described in chapter 4, tourists also displayed rude and ignorant behavior. Lisa, the British volunteer, reflected on her work handing out samples outside the chocolate store:

> Some tourists are just so unbelievably rude. I remember one lady – I was just asking if she liked a chocolate sample, and I mean I'm really smiling and friendly and all, and she basically just put her hand up in my face and said, "Go away, go away!" And the guy who dresses up as the Inca was standing near me, and as she was walking away, he was like, "Fucking tourists!" – as if to say, "Look, don't worry, hon. They're all like that sometimes."

Resistance from locals includes examples of both veiled and open resistance. By indirectly insulting the tourists, "the Inca," introduced in chapter 2, expressed solidarity with a foreign volunteer. On other occasions when he chatted with me, he openly mocked tourists, causing me and vendors nearby to laugh. Humor is a common strategy of veiled resistance, which I saw employed frequently. Johnny often imitated tourists behind their backs, and so did his colleagues. One evening, María told me about a couple from Lima staying at her guesthouse. "My God, that woman complained about everything. She didn't like the toilet paper; she wanted the softer kind. But I said, 'Sorry, we are in the mountains here; we only have this kind.'" I heard her tell this story a number of times to friends and neighbors, who usually joined her in laughing heartily. As addressed earlier, people in the highlands often feel discriminated against by Peruvians from the coast, and giving misinformation about the availability of toilet paper and making fun of them was a means of pushing back. Similarly, Jill Sweet's research among the Pueblo people showed that burlesquing White tourists was a common strategy of resistance (1989).

Sometimes locals straddle the line between being polite and telling tourists off openly. Naida recounted the following incident:

> One time there was a group of Argentineans and Chileans here at the same time, and the Argentineans said to throw out the Chileans, and the Chileans wanted me to throw out the Argentineans. I said, "Please, if you have to fight, you can do that on the street but not in my house. You Chileans took some of our land, and you Argentineans should have supported us in the war against Ecuador, but you helped them. I should be mad at both of you. But since I'm polite, I welcome you all in my house, so let's be calm and get along." Then they were quiet and went to their rooms.

By remaining polite and using a broader political perspective to appeal to their guests' understanding, Naida managed to get them to comply. On another occasion, she more openly told someone off: "There was an Argentinean hippie – I think she was high – and she was throwing garbage into the street. I said, 'Hey, you can't just throw garbage here. Take it with you.' And she said, 'Show me some respect here.' And I said, 'What? It is you who is not showing respect. You are in another country. If you want to throw garbage around, stay in your own country!'" Similarly, an older guide who had worked in the

Cusco area for many years recounted his experience with a French tourist, imitating her with a high-pitched whining voice:

> There was this short, old French woman who complained about everything on the trip. *This* was bad, *that* was bad – *everything* was bad! Then at the end, she suddenly wanted to give me a tip and recommend me to her friends. I didn't accept the tip and said, "Please don't recommend me to your friends; I do not want to travel with more people like you!"

Local people clearly find different ways to resist and challenge tourists' behaviors. Most of these acts consist of veiled resistance and are unlikely to alter the larger inequalities. Open aggression against tourists was very rare, and during my stay, I heard of only one specific case in which a tourist was mugged, a case that locals described as very unusual and worrisome. However, tourists are still vulnerable. Drawing on Foucauldian notions of power as always shifting, So-Min Cheong and Marc Miller emphasize that visitors are often unable to communicate in the local language, unfamiliar with cultural norms and political realities, and often highly visible as outsiders; thus, they occupy "insecure positions" and can easily become targets (2004, 380–3). As outlined earlier, female tourists are generally more vulnerable than their male counterparts, and they may be criticized far more than males for transgressing social boundaries, which can lead to violent backlash (Frohlick 2010).

Hostilities against tourists are also expressed in the realm of the supernatural. Across the Andes, stories are told about the mythological figure of the *pishtaco*, a large White man said to kill indigenous people to use their body fat (Hill 2008, 260; Weismantel 2001). Weismantel analyzes the *pishtaco* stories as reflecting an underlying economic reality: since colonial times, White foreigners have been extracting the life force, in the Andes often depicted as fat, from local people, and the stories highlight the extreme brutality of these interactions. While the *pishtaco* of a hundred years ago was said to use body fat for candles or as holy oil, according to contemporary versions, the fat is used for facial cream or machine lubricant. The modern adaptations reflect the continued relevance of this story, and in Ecuador, several tourists were attacked because locals considered them to be *pishtacos* (Weismantel

2001, 203). This type of local resistance points to the darker side of tourism encounters and to the importance of acknowledging its continuities with colonialism. Soledad, a local woman, told me about a recent case near Ollantaytambo. As part of her work with an NGO, she was visiting a small highland community about half an hour's drive above the valley. At the end of the day, she spoke with local men about arranging a ride back. When a tall, blonde foreigner walked by, the men looked suspiciously at the unaccompanied woman and agreed that she could well be a *pishtaco*; they then debated whether to get sticks and hit her. Soledad intervened and managed to convince them not to do this. It is usually men who are suspected of being *pishtacos*, but the foreigner's gender may not have been immediately obvious to the locals; walking by herself, probably wearing pants and hiking boots, she did not act in accordance with local definitions of femininity. The figure of the *pishtaco* reflects ongoing tensions based on the history of colonialism and exploitation by outsiders, and this colonial legacy can result in real danger for foreigners.

Another instance revealed aggressive attitudes toward tourists. Over a shared dinner, Valentina commented on the tourist dangers in Juliaca, a border town with a rough reputation. "I've heard that tourists can't even go down the road by themselves. They have a lot of gold there, and sometimes they kidnap tourists and kill them. You know why? To use as offerings in the mines!" When I questioned her further, she said she had heard this from different people. "That's how it is," she added matter-of-factly. I doubt we can take the story at face value, but the way it was told is revealing. Neither the idea that a ritual offering could include the killing of people nor the fact that tourists were the reported targets seemed especially noteworthy or out of the ordinary to Valentina. It suggests that sentiments of resentment and aggression toward tourists, whether only felt or actually enacted, may be more commonplace than people generally acknowledge.

The majority of people in Ollantaytambo considered tourism a positive force and wanted it to develop further. As discussed in the previous chapter, many women benefit from flexible and low-stakes work opportunities that can be combined with domestic duties. However, these changes are accompanied by growing material inequalities and responses of open and veiled resistance. How effective are these strategies? Michel Foucault considers resistance as inherent in power relations but also often transient and short-lived (1978, 95–6). While in many instances local people openly expressed frustration and aggression toward tourists and foreign residents, these demonstrations did not seem to have long-term effects. Víctor, Naida's husband, commented that one had to be really careful because when tourists write a bad online report it can damage business. Though at times very outspoken, María told me that often it was not worth the trouble to argue with guests, and instead she preferred to send them away with a refund.

As material inequalities are growing in the formerly more egalitarian community, we find continuities of cooperation as well as conflict. Acts of resistance, including supernatural beliefs and practices, can be seen as leveling mechanisms aimed at "taking down" those that have distanced themselves through greater wealth. As described by Edward Evans-Pritchard, the practice of witchcraft can convey messages about accepted behaviors and serve to restore social balance (1937). Gender is also an important factor. In the cases of physical threats and violence described above, the aggressors were all local men; their targets were men from outside the community or the country who had moved in and established themselves successfully. In Colombia, deregulated market conditions and insecure employment have been described as "destabilizing factors for masculinity ... conditions [that] erode one of men's main identities, that of the provider" (Gómez Alcaraz, Hernán, and García Suárez 2006, 103). Based on the principle of "limited good" (Foster 1965), the newcomers were likely considered competition for scarce resources as well as a threat to local masculine identities. Local men's open and veiled acts of aggression can be seen as attempts to reclaim their position of dominance. While I did not witness open violence against foreign women, I heard several locals refer to Jacquie as a witch, which may have been prompted by envy of her successful hostel. Also, if local men feel that their gender identities are on the line, successful women may be perceived as even more threatening than other men.

Being more connected to the community than tourists, foreign residents also become more dependent and consequently more vulnerable to acts of hostility. A few months after the black magic incident, Ben and his girlfriend experienced growing health and relationship troubles; eventually, they broke up and returned to their home countries. Local friends have since taken over the guesthouse. Much later, Ben's partner told me, "When I think back to that time it all seems crazy. But that was our reality then." As another foreign resident, I generally felt well accepted, yet locals' responses to my insomnia also showed that my privileged position warranted envy and hostility. Traditional networks of support and cooperation persist but are increasingly stretched in the face of rapid economic changes. The situation is similar in Costa Rica where members of the middle class and foreign nationals benefit more from tourist developments, and although women gain more power, class differences increase (Ferguson 2011, 366–8). Expressions of black magic to mitigate growing inequalities remain largely hidden from tourists, but in the next chapter, I explore how certain forms of spirituality become commodified for tourism and can also function as strategies of resistance.

6

Marketing Spirituality and Romance

One of the local men I came to know well was José; he had grown up in Ollantaytambo and now ran a guesthouse and organized spiritual tours to nearby Inca sites. During most of my stay, I saw him busy renovating parts of the house and reconstructing an Inca stone wall. One afternoon, I walked past with Naida, who turned to me and said jokingly, "And here we have the last Inca." On other occasions, he would stop his work, wipe the dirt and sweat off his face, and chat with me. Several times, he pointed out the shapes of the large stones making up the bases of houses. "Every stone has its meaning," he said. "Many people here don't know, but there are special figures in the rocks; the Inca put them there. My grandfather told me all about this when I was a little boy. I didn't value it then, but as I grew older, I could feel it more. It is about feeling it, not just knowing. You need to feel the energy."

I first met Giulia as she was chopping vegetables for dinner in the guesthouse kitchen. A German-Italian woman of about forty, she had her curly brown hair tied back with a scarf and exuded an easy-going confidence. Giulia had been on extensive solo trips across Europe, Asia, and Latin America and financed her travel by selling handmade jewelry and taking odd jobs.

Most evenings she spent at the local campsite, the cheapest place in town, which was also frequented by local young men who came for evening parties and sometimes set up camp for a while. "There is this guy I like," she told me cheerfully.

"I sat next to him in the sweat lodge. I would like to, you know, exchange some energy with him."

Apart from environmental and cultural attractions, Peruvian tourism has developed a distinct form of New Age tourism, locally called spiritual or mystical tourism (*turismo místico*). The term New Age broadly refers to an eclectic spiritual or religious movement that arose in the 1970s in the West and combines elements of Eastern philosophy, Western esotericism, and indigenous beliefs and practices (York 2001, 363). Central themes are the interconnectedness of all beings, the divine nature of humans, and the move toward a major shift of consciousness or New Age (Granholm 2013, 60; York 2001, 364). In the context of Peruvian tourism, especially in the Cusco area, we find a blending of these global New Age themes with distinctly Andean traditions (Hill 2008, 252). Similar processes of syncretism have been described in the marketing of Mayan sites in Mexico (Castañeda 1997), and Sagar Singh has argued that tourism is inherently spiritual, involving focus on the self as well as "self-transcendence" (2019, 84). Tourists can purchase experiences, such as the participation in ceremonies, as well as material goods like crystals and figurines said to hold spiritual power or positive energy. New Age discourse and practice permeate regular tourism, and several tour guides, who did not call themselves spiritual guides, told me that they frequently spoke about the mystical powers of Inca and pre-Inca sites and performed ceremonies for tourists.

In addition, spirituality and feelings of love can be closely intertwined (Singh 2019). In the Cusco area, we find that spiritual tourism is strongly entangled with the marketing of romance, and local men effectively draw on New Age themes to appeal to foreign women. In places like Hawaii (Desmond 1999) and the Caribbean (Kempadoo 2004; Pruitt and LaFont 1995) local bodies are quite explicitly romanticized and sexualized for tourism. Tourist imaginaries are always driven by a desire for what is different, and often this can converge in a fascination with the exotic Other on a personal and sexual level, as Lorraine Williams describes in her book on sex tourism in Brazil (2013). Sex tourism is primarily associated with traveling men seeking local women, and while male tourists may be criticized, their behavior is often naturalized as part of the adventure discourse and normative masculinity (Enloe 2000; Frohlick 2010, 56). In the Andes, sex tourism catering to visiting men is not nearly as prevalent as in other parts of Latin America; rather, it is often local men who seek out relationships with foreign women who can provide them with sex and often substantial material support (Babb 2012; Bauer 2008; Wilson and Ypeij 2012). Partners usually speak about their relationships in terms of romance and emotional connection rather than just sex, fitting with what Deborah Pruitt and Suzanne LaFont have defined as romance tourism (1995, 423). However, others have argued that people's motivations are varied and shifting and that it is more useful to understand sex tourism and romance tourism as a continuum (Herold, Garcia, and DeMoya 2001). I will explore this distinction further after first outlining the main themes of New Age discourse and the ways in which these are used. I consider issues of commodification and cultural appropriation and examine who controls these practices and who benefits. Both spiritual and romance tourism involve commodification and selling out to foreign demands, yet the narratives produced can also serve as strategies of veiled resistance and allow local people to challenge unequal power relations, at least temporarily. Spiritual tourism involving the use of hallucinogenic plants, notably ayahuasca and San Pedro, has expanded greatly in recent years (see Fotiou 2016 and Prayag et al. 2016) but is not included in my discussion here.

SPIRITUAL TOURISM AND NEW AGE DISCOURSE

Initially, New Age tourism was not an explicit focus of my research. However, over the course of my time in Ollantaytambo, I was struck by how frequently certain themes came up in conversation with both locals and visitors, so I decided to pay more attention to this. In my approach to discourse, I follow Michel Foucault, who argues that the ways in which people consistently speak about a certain topic can, over time, constitute reality. Discursive practice is thus not just a transcription or evaluation of an outside reality but rather produces it (Foucault 1972, 48–9). Discourse also functions as a main avenue of producing and solidifying power relations (Foucault 1972, 1978). In the context of tourism, we can see how certain intangible tourism imaginaries can become manifested over time (Salazar and Graburn 2014, 2). While New Age discourse varies widely, we can identify key themes and analyze how they function for the different participants involved (Granholm 2013, 52). I will do this for the following four topics that have emerged from my research: the strong romanticization of the Inca and contemporary indigenous people, the call to share and appropriate traditional practices, the focus on personal destiny, and the value placed on feeling over thinking.

First, New Age discourse selectively draws on highly positive aspects of Inca and indigenous culture, resulting in narrow and essentialized presentations that appeal and can be marketed to foreigners. This selective focus dovetails with *incanismo*. As described earlier, this political movement was largely driven by the urban elite who, while extolling the virtues of their Inca ancestors, sought to set themselves apart from contemporary indigenous people whom they regarded as inferior (Hill 2008; Silverman 2002; van den Berghe and Flores Ochoa 2000). Likewise, tourist guides today will describe the Inca state as a model empire and the Inca people as highly spiritual and sometimes superhuman. For example, guides in Ollantaytambo highlight the ingenious move of Inca leader Manco Inca, who overpowered Spanish forces by flooding the plain through water channels in the sixteenth century. Native tour guides in Alaska use similar strategies by emphasizing the Tlingits' win in a battle against the Russians (Bunten 2008). Both for the Inca and the Tlingit, victory was short-lived and did not halt the invasion of colonial forces; however, these narratives serve to counter the stereotypes of indigenous people as victims and make a political statement about their endurance and strength. Similarly, when Diego explained Ollantaytambo's main Inca site to me, he emphatically stated, "Everyone calls these places ruins, but that's not right. For us these are sacred places; they are not ruined." Highlighting his people's ongoing relationship with the site, he takes an active part in "the process of heritage-making," similar to what Ronda Brulotte (2017, 57) describes among vendors at Mexico's World Heritage Site Monte Albán.

However, while the historical Inca are glorified, their contemporary descendants are often treated in a derogatory manner. Guides on the Inca trail may tell tourists that the *campesinos* they encounter are not originally from the area and thus not authentic (Ypeij 2012, 26). As cultural brokers, guides hold power in shaping visitors' impressions; thus, their often disparaging attitude toward more indigenous people can affect tourists' impressions. In Ollantaytambo, I have frequently heard people describe members of nearby highland communities as "dirty." Plus, just as in the *incanismo* movement, the very connection between the Inca and modern indigenous people is sometimes called into question. A professor in Cusco considered the Inca a "different and noble race" that fell victim to disease, while present-day indigenous people, whom he referred to as "dirty thieves," were supposedly of largely Spanish descent. A similar dichotomy is vividly illustrated in the film *Incidents of Travel in Chichén Itzá*, which shows how predominantly American New Age seekers venerate aspects of classic Mayan culture while expressing disregard toward the contemporary population (Himpele and Castañeda 1997). In Peru, along with sidelining modern *campesinos*, many people also emphasize the similarities between the Inca and contemporary Westerners. For example, the stone steps in many Inca sites are quite high, and guides are quick to point out that "the Inca were tall, just like you." In the context of Inca architecture, social welfare, or astronomy, a comment frequently heard is that "the Inca were very advanced, like you." In these ways, elements of the past are mobilized in discourse to present a more empowering version of history, a process of actively constructing the cultural "heritage-scape" (Di Giovine 2009).

While many urban and *mestizo* Peruvians seek to distance themselves from indigenous people, foreigners are generally fascinated by them, and mystical tourism produces essentializing narratives catering to these tourist imaginaries. In their analysis of postcards from Cusco, Aviva Sinervo and Michael Hill note the emphasis on indigenous people's spiritual nature; they argue that "foreigners thereby avoid the issues of material inequality, taking refuge in the idea that cultural and spiritual riches make Peruvian children – indeed, indigenous Andeans more broadly – a happy people despite their suffering" (2011, 134–5). One morning, I was chatting with Jacquie outside her hostel. A group of young children from a nearby highland community walked past us, all dressed in traditional handwoven clothing.

"Look at them; aren't they beautiful?" Jacquie said. "They are pure spirit." Her comment was expressed with warmth and admiration, and the fact that as a long-term resident she frequently interacts with indigenous people places her comment in a different category than a tourist's. However, this description also has problematic implications. Calling the children "pure spirit" elevates them above other people but thereby also stages them as Other. This distancing can provide comfort; the children are so different and superior to us that we do not have to confront the blatant material inequalities that separate us.

A second common New Age theme consists of the call for broader sharing of traditional cultural knowledge. Imasmari, a spiritual organization based in Cusco, produced a three-minute video that provides a good introduction to Andean New Age themes (Werber 2016). The clip shows predominantly young White people performing ceremonies in Inca sites, while people with indigenous Andean characteristics are featured only briefly. The items used in the ceremonies range from crystals and prayer beads to Tibetan singing bowls, and someone is wearing what look like deer antlers. This scene is again reminiscent of the situation in Chichén Itzá where Western New Age seekers have invaded Mayan temples in order to perform their own spiritual ceremonies (Himpele and Castañeda 1997). The voice-over describes the purpose of Imasmari as follows:

> Five hundred years ago, a group of Incan people retreated to the high Andes mountains to preserve a prophecy. The prophecy states that we have now entered the dawn of a New Age, in which humans will remember their true essence. Though in order for this New Age to manifest, there must be a rejoining of the paths of wisdom that were long ago sent to the four ends of the earth. This great reunion … will catalyze a greater shift for a consciousness of grace. We believe this is why so many people from around the world are now drawn to the Incan capital of Cusco, Peru. As people of all lineages and nations cross paths here, *we* are putting the wisdom pieces back together; *we* are the great reunion.

It is worth noting that the shift of consciousness is described as contingent on people from different parts of the world coming together to learn and share. Indigenous Quechua beliefs are subsumed under a larger New Age movement; they are de-ethnicized and de-territorialized and thereby made accessible to outsiders (Hill 2008). Likewise, at a spiritual retreat center near Cusco, the American owner stated that, in response to growing global problems, Quechua elders have chosen to teach sacred rituals to noncommunity members and foreigners (Gómez-Barris 2012, 75). This notion evokes common tourist imaginaries, but these are characterized by underlying power inequalities (Salazar and Graburn 2014, 876). In mainstream Peruvian society, Quechua people have been systematically marginalized and discriminated against. While the mystical tourism sector positions them as the holders of spiritual authenticity, indigenous people usually do not have the cultural know-how or language skills to market their traditions, so they benefit little or not at all. Commodification, "a process by which things (and activities) come to be valued primarily in terms of their exchange value, in a context of trade, thereby becoming goods (and services)" (Cohen 1988, 380), has been shown to result in cultural loss (e.g., Greenwood 1989). One of the most common ceremonies performed for tourists is the *despacho* or offering to Mother Earth. Traditionally, the ceremony involves a healer assembling a number of objects, usually coca leaves,* sweets, and sometimes confetti or flowers; wrapping them in paper or cloth; and then burning or burying the offering. At the retreat center in Pisac, the *despacho* ceremonies were conducted in under an hour, compared with indigenous versions that typically include music and dance and last for several hours. Macarena Gómez-Barris describes how the "engagement with local culture and religion was replaced by ethnicized spectacle"; the American and a *mestizo* man directed most of the ceremony while the Quechua man was delegated to the role of "authenticating figure" (2012, 75). Like most Andean ceremonies, *despachos* are traditionally a communal affair and seek to maintain or restore balance between people and the land (Bolin 1998; Greenway 1998). Images on the center's website, however, show elaborate individual offerings (Paz

---

* Coca (*Erythroxylum coca*) is a bushy plant cultivated in the lowlands. It has been much maligned because it is the basis of cocaine. However, in the Andes it has been used for centuries, where it figures prominently in rituals and, when chewed, provides high levels of vitamins and minerals (e.g., Davis 1996).

6.1 *Despacho* prepared by Alberto and Diego, by Karoline Guelke

y Luz 2017). Whereas many locals of Cusco emphasize the concept of *ayni*, or reciprocity, between people and the environment, foreign spiritual seekers often speak about personally receiving positive energy from specific places and ceremonies (Hill 2007, 453). *Despacho* ceremonies are often commodified for the tourist market, and while indigenous people may be invited, they arguably have little control over the process.

However, this was different in some of the *despacho* ceremonies I witnessed and heard about in Ollantaytambo. During the winter solstice (June 21), I joined Diego and Alberto for a *despacho* for tourists. We camped on a hillside above Ollantaytambo, and at sunrise, the brothers prepared the offering shown in figure 6.1, which included bread, flowers, chocolate, corn beer, and coca leaves. We each received a set of three coca leaves to use as an offering, and Diego poured the corn beer in the direction of the rising sun. He explained that the purpose was to honor the earth and sacred mountain peaks, reflecting an emphasis on the rite as maintaining a connection with the environment rather than just on benefitting us as individuals.

Alberto also shared photos of a *despacho* for the photovoice project, explaining that he sometimes invites an elder from the Q'ero community to perform the ceremonies (see figure 6.2). While he felt confident performing the ceremony himself, he said it was better that a *curandero* (healer) did it. In this arrangement, Alberto organizes the logistics and functions as the cultural broker who provides translations and explanations to the tourists. His comments indicate more of an equal partnership in which he acknowledges and respects the *curandero*'s expertise and gives him control over when and how to perform the ceremony.

Two other tour guides I interviewed regularly performed *despachos* themselves. When asked about the meaning this had for them, one explained that his family was Catholic but that he had been interested in learning his traditions and enjoyed doing it. Similarly, Amy Cox Hall describes how guides on the Inca trail learn about "being Inca" (2017, 42). The other guide mentioned that his family performed their own *despachos*, so he was familiar with the practice and found he could make good money doing it. It does not seem that the ceremony has become a meaningless exercise only. While the open sharing of indigenous knowledge can be problematic, performances for tourists can also contribute to the revival and validation of cultural practices.

A third common New Age trope is that of destiny and personal calling through which foreigners explain their presence. In my interviews with tourists, these themes emerged repeatedly. People spoke about serendipitous ways in which they ended up in Ollantaytambo and felt they were meant to be there. An American woman in her fifties, who described herself as a spiritual seeker, often commented on the role of destiny. She spoke of "knowing on a deep level" that she had to come to the Andes and "feeling called" to learn and share her knowledge of spirituality and healing. Several travelers who did not identify as particularly spiritual used similar phrases. A young Swiss backpacker traveling with her boyfriend described that upon their arrival in Ollantaytambo, they felt that "it was just the right place, like we were meant to be here." Similar parallels between backpacking ideology and New Age thinking have been identified elsewhere (Binder 2004).

6.2 Q'ero elder performing ceremony for tourists, by Alberto

Traveling to another place, of course, involves paying for accommodation and food, and while many backpackers manage to spend little money, the fact remains that they can only follow their calling because they have the resources to do so. According to Michael Hill, "New Age ideologies about predestined spiritual meanings of serendipitous occurrences reframe the potentially alienating dimensions of commodification and inequality by imparting sacred meaning to capitalist exchange" (2008, 268). Macarena Gómez-Barris has identified this theme prominently in the writings of Diane Dunn, the owner of the retreat center in Pisac, who talks about her "deepest calling," which led her to Peru, thus justifying her right to practice Andean spiritual traditions and build up her retreat center there. She is one of many foreigners who have bought land in the area, leading to land scarcity and rising property values (Gómez-Barris 2012, 76).

Last, the incident of José instructing me to feel the energy in the rocks reflects a fourth major trope: the value placed on the process of feeling over thinking, or "heart knowledge" over "mind knowledge" (Gómez-Barris 2012, 74). As Jacquie explained,

> The more tourists can ask for the local people to teach us, you know, the more it benefits both the tourists and the local people, and so that is something I have promoted in my hostel, always recommending that my guests go with local guides, even if they don't speak much English. I mean, if they don't understand everything, it's more about sinking into their hearts, because this is a heart-centered culture, and so just feel how he connects to the land, how his energy and his presence is. And so that my local buddies who take the tourists out feel good about themselves, because they can share something that is important information in the world, how to be in your heart, how to connect to *Pachamama* [Mother Earth].

Both José and Jacquie remind us of the limits of rational "mind knowledge" and point to another important avenue of connection and understanding. Indigenous approaches to learning have long included the heart dimension as a prominent aspect (see, for example, Bopp et al. 2004; Cajete 1994). However, while this is an important teaching, it can also be used in a way that restricts certain understanding. In her analysis of New Age ceremonies, Macarena Gómez-Barris argues that when visitors are simply asked to feel instead of think, the cultural context of the teachings is erased, and opportunities for a more nuanced understanding of indigenous peoples are wasted (2012, 74). In addition, many craft and jewelry vendors use a similar discourse, and the emphasis on feeling and positive energy certainly functions as an effective marketing strategy. I cannot count the times that I picked up a figurine or a piece of jewelry only to hear the vendor tell me that this would attract positive energy, a concept as ill defined as it is powerful, and something that had to be felt. This compelling narrative often lands the vendor a good sale and gently undermines the potentially higher status of foreigners based on their higher levels of formal education. The argument that "heart knowledge" is superior to "mind knowledge" challenges that power hierarchy; it positions local people as the ones who know how to access this kind of important knowledge and Westerners as having to learn from them.

The expressions and effects of the Andean New Age movement are complex. Many Westerners are driven by a sincere desire to find healing, happiness, and guidance for life (Ghasarian 2014), and simplified ceremonies, material goods, and traditional knowledge are marketed to meet those needs. While this leads to problematic romanticization, commodification, and cultural appropriation, New Age tourism and discourse are also tools that can be used on a personal and political level. As Florence Babb reminds us, "we need to view tourism as far more than selling out to global interests and to understand it as fundamentally linked, for better or worse, to the refashioning of histories and nations" (2011, xiii). New Age discourse can allow for a reframing of a violent past characterized by exploitation and discrimination. As in many parts of the world, in the Cusco area colonial heritage sites have become tourist attractions, yet in people's self-representation, Inca identity figures more prominently. By meeting foreign visitors as a descendent of the spiritually superior Incas, rather than of an exploited and colonized people, Peruvians assume a more powerful position vis-à-vis tourists, and I heard several local people state that they did not want to be portrayed as victims. Success as a guide can also bring social status. British expat Kate commented on the pride she could see in two young sisters when they spoke about how Westerners admired their father, a spiritual tour guide. Jill Sweet's work with the Pueblo people illustrates how choosing when to share or withhold information allows local people to exercise significant power in their interactions with visitors (2010, 142). Considering most tourists' fascination with the topic, this may apply even more to knowledge about spiritual beliefs and practices.

With regards to gender, it is important to note that most spiritual guides, and most guides in general, are men (Ypeij 2012, 25), and spiritual tourism appears to be especially attractive to Western women. Considering this, New Age tourism brings greater benefits to local men than to women, both in terms of material profit and increased social status. It has been shown that, as employment and state support become more precarious in a neoliberal climate, men often struggle to fulfill their role as provider, which, as discussed earlier, can result in violence and conflict. By using their cultural capital, and often appropriating that of other groups, men can find new avenues of work in the New Age sector of tourism. As outlined previously, local women can benefit from small-scale businesses, but they do not participate equally in mystical tourism. Local men also employ New Age tropes to establish relationships with foreign women, adding another gendered dimension, which we will turn to next.

Marketing Romance

In the Cusco area, local men who look for relationships with foreigners are called *bricheros*. This term possibly derives from the English word breeches, referring to the three-quarter-length pants traditionally worn by indigenous men, or, more likely, originating with the term bridge; thus, the *brichero* (or the female version *brichera*) is someone who builds a bridge between two peoples and cultures (Bauer 2008, 613; Ypeij 2012, 27). American expat Jacquie referenced the concept of the bridge: "*Bricheros*, yes, I think it is a really positive way for the connection, the bridge between the worlds, and, I think that's a lot of the tourism here, women coming here and finding these cute young men." It is important to note that, in contrast to this optimistic assessment, many local people use the term *brichero* in a critical and derogatory manner, and the men themselves usually do not self-identify as such. Yet, given that the term is so commonly employed by both locals and researchers, I chose to use it as well – keeping in mind the complexities and varying connotations of this label.

Ollantaytambo is far smaller than Cusco, but one can easily find men who fit this description. The most visible examples are the young craft vendors who sell their goods in the street below the main archaeological site. Many make their own jewelry or resell what they purchase elsewhere. A popular spot is by the two large Inca stones that sit in the main street; often the vendors lay out their wares on top and sit nearby working, chatting, and calling out to potential buyers. Apart from street vendors, tour guides are often associated with the role of the *brichero* (Ypeij 2012, 27), as are men working in other parts of the tourism sector. I begin by examining how *bricheros* use New Age discourse and then consider the motives and power relations involved. My findings indicate that, while in many ways these relationships can be mutually beneficial, foreign women tend to hold more of the power, thus replicating the global power inequalities we so often see in international tourism encounters.

BRICHEROS AND NEW AGE DISCOURSE

The strategies men use to connect with foreigners draw heavily on *incanismo* and the New Age discourse described previously. In response to the romanticization of Inca and indigenous people, *bricheros* tend to cultivate a look that is distinct from local men, often growing their hair long following Inca tradition and wearing jewelry with Inca symbols (see figure 6.3).

In conversation, they claim a close connection to their Inca ancestors as well as to *Pachamama* (Mother Earth). At the same time, they usually acquire fluency in English and broad cultural knowledge so as to converse with foreign women on a number of topics. In her analysis of cross-cultural romance in literature, Mary Louise Pratt has observed that colonial love interests are usually not completely non-White but often of mixed ancestry or with strong European affiliations (1992, 100). Similarly, *bricheros* cultivate a composite identity, simultaneously the exotic Other and somewhat familiar. The resulting persona clearly appeals to Western women, as indicated by the comments of several female tourists. Sara, a Swiss backpacker, and I had spent

6.3 *Brichero* craft vendor with tourist, by Karoline Guelke

most of the day hiking to a smaller Inca site, and we passed the street vendors on our way back. Looking at one young man in particular, who was perched on one of the massive Inca rocks, his long hair in a ponytail and his bare torso covered by only a vest, Sara turned to me and whispered, "Wow, he looks like a real Indian – so cute!" On another occasion, I shared dinner with two middle-aged American women. One of them commented on the "beautiful Indian guys with their long hair" while the other agreed enthusiastically: "We had just a beautiful guide in Cusco; he looked like a real Inca, very charming. He told us all about how he grew up in the mountains, and he still speaks Quechua." While the men are not as overtly sexualized as in, for example, the Caribbean (Kempadoo 2004) or Brazil (Williams 2013), they are clearly objectified and racialized. Women respond strongly to their physical appearance, particularly their "Inca look," and the men's stories of Inca and indigenous heritage increase this appeal.

As discussed, speaking of destiny and personal calling allows foreigners to justify their presence and sideline inequalities, but the trope is also employed effectively by local people. In Mario Guevara Paredes's short story *Gringa Hunter,* a *brichero* in Cusco describes his strategy for luring in a Western woman as follows: "I used an old trick with the *gringa* that's always worked. It was to convince her that our meeting wasn't by chance, but rather it was due to the magnetism this city radiates, making it possible for us to meet" (1998, 101–2). As a single Western woman spending a lot of time walking around and sitting in cafés by myself, I experienced the *bricheros'* strategies first-hand. While I was sitting on the stone steps watching the goings-on in the plaza one afternoon, a young man sat down next to me and declared, "I think it was destiny that we met!" He then tried to convince me to go hiking, as he wanted to show me a supposedly unknown Inca site nearby. Framing a chance meeting as destiny can serve to hide other possible motives for seeking a connection, such as sex and material support, making it more difficult for women to reject the men's advances.

Another common New Age trope employed by the *bricheros* is reflected in the offer to share special knowledge about Inca ancestry and spirituality. Mario Guevara Paredes's narrator explains his epistemological approach to impress a foreign woman: "I had another way of perceiving reality. And it wasn't the simple reality the majority of people see ... because my perception came from an ancient belief that only belonged to the chosen ones" (1998, 102). The type of perception he describes is clearly differentiated as belonging to a select few. Positioning himself as the carrier of secret knowledge results in a mysterious and desirable persona and thus provides a greater power base from which to interact with foreigners.

Furthermore, it is not difficult to imagine how the argument of focusing on feeling over thinking can be used as an effective strategy for romantic advances that can lead to exploitative situations. Stefanie, a German backpacker in her thirties, recounted an experience of sexual abuse in nearby Bolivia. She had arranged for an Andean cleansing ritual and described to me how, as the ceremony went on, her guide had started touching her while keeping her captivated with his explanations:

> He had such a good story. It was like, I couldn't step out of his story, you know? About the energy and how I just needed to open up to it and *feel* it. And it was only to get me to have sex, and after it was over, I thought: What the hell was that? I felt guilty; I felt ashamed. I think his story was just about getting closer physically, sexually, but at the moment I couldn't see that.

Female travelers are at greater risk of being sexually exploited than males (Frohlick 2010; Jordan and Aitchison 2008), and the trope of "don't think, just feel" can be used against them. It should be noted that it is not just about feeling anything but rather the call to feel something specific. Stefanie said she wished she had *trusted* her true feelings and left. She was not encouraged to get in touch with what she was actually feeling but rather pushed to feel something in particular, and in this case, to even ignore and shut down part of her emotions, which left her highly vulnerable. Stefanie's story was the most explicit description of sexual exploitation I heard from tourists, but warnings about male guides were common, both from tourists and local people.

However, many female tourists also commented positively on the theme of feelings. Nina, a Dutch woman in her twenties with striking blue eyes, told me that she had taken time off from her busy work schedule to spend two months traveling around South America. With a somewhat flexible schedule, she had decided to extend her stay in Cusco in order to spend three weeks with a local man. "I was so stressed when I came, from work and all that – so much in my head all the time. But this guy, he showed me how to take it day by day, how to feel more. I think I really needed that." It is interesting that her description draws on the exact same themes used by the men, that of a Westerner who is thinking too much and needs a local person to get in touch with her feelings. Similar explanations have also been documented among male clients of Brazilian sex workers (Williams 2013, 85), reminding us of the difficulties of separating sex tourism from romance tourism.

MOTIVES AND EMOTIONS

Mario Guevara Paredes's narrator describes his strategy of speaking about ancient Andean teachings that include the "overcoming of one's affections for material things" (1998, 102). This is another familiar New Age trope that appeals to Westerners and also gently undermines their power position based on wealth. In addition, it serves to conceal the irony that for many *bricheros* material support is one of their main objectives. Giulia, described earlier, had started a relationship with the man from the campsite; they traveled to different parts of the valley, and she casually mentioned that she was paying for everything. "It's okay," she said, "I don't have much money, but I'm better off than he is." This statement describes the common situation of material inequality between foreign women and local men, as well as the general willingness of the women to pick up the tab, whether for short-term excursions or more long-term ventures. When I asked Mauricio, a local tour guide, about his relationship with a Canadian volunteer, he laughed and said, "She's great. We are having fun – and it helps with the money." Johnny, the artist and street vendor I often talked with, repeatedly joked, "*Señorita*, bring me a girlfriend from Canada, please! Just bring her in your suitcase!" When I asked him why he wanted a foreign girlfriend instead of a Peruvian, he replied that local women were troublesome and that he wanted to go travel and see the world. It is a realistic assessment that likely only a foreign partner could afford him this opportunity. Women supporting their male partners financially has been reported from other places, such as the Caribbean (Herold et al. 2001; Kempadoo 2004), Costa Rica (Frohlick 2016), and Bali (Virmani 2009).

Another clue about material motivations is the fact that cross-cultural relationships are not just initiated by the partners themselves. Jason, the American adventure guide, recounted an occasion when local people tried to set him up. While he was eating breakfast at a local restaurant, two elderly women questioned him why he did not have a Peruvian girlfriend. One of them pulled in her reluctant granddaughter, who, according to Jason, looked about fourteen years old. This incident illustrates that relationships between foreigners and locals are not simply rebellious choices of the young that are regarded as improper by the older generations, though I have heard of such cases. In this situation, the two older women clearly had an interest in forging a cross-cultural relationship. While we cannot know their exact motivations, it seems likely that they were materially based. Jason was successful in his work and very well-off by local standards, which would have benefitted the woman's granddaughter and extended family members as well. Although this example is the reverse of the typical *brichero* situation involving local men and foreign women, it nevertheless provides clues as to why local people may seek relationships with foreigners. In addition, the association with foreigners can provide some access to cosmopolitanism, which is usually reserved for wealthier people, as Williams describes for female sex workers in Brazil (2013, 134).

For foreign women seeking local mates, the primary motivation seems the appeal of an exotic partner as part of the travel experience. Jacquie described her first visit to Ollantaytambo as follows:

> When I first landed here I went to hang out at the bar every night. It was my midlife adventure; ha, it was awesome! It was me and all these men, all these beautiful young men! And they were just like – woosh – you know, coming on to me. I didn't really understand what they were saying, but I loved the attention and the energy ... just this crazy amazing focus and attention that made

> my feminine go whohoo! The Western men, you know, they don't really know how to be a man; they are all confused.

While I did not ask specifically about sex, several women alluded openly to it. Jacquie often mentioned her sexual experiences in Peru, and Giulia joked about "exchanging energy" with the man she had met. At the local bars, one can see couples touching and kissing. But apart from material benefits for the men and sexual experiences for both partners, these liaisons can have other benefits as well.

A cross-cultural relationship can mean social capital for both partners. It seems that simply being seen with a foreign woman in public can increase a local man's status, especially among his peers. Lisa, the British volunteer, described an experience at a festival, where local men had persistently asked her to dance:

> I get the feeling that it is just because I'm a *gringa*. Every five minutes it was like – I hadn't learned by then to say no – so every five minutes I was dancing with somebody. And while they were dancing with me, they would kind of be looking around to see if their mates would see them dancing with a *gringa*.... It wasn't because they just wanted to dance with me or thought I was pretty, you know? Sometimes it really felt like they were like, "Oh look at me, I'm dancing with a *gringa*!"

Just as for local men being seen in public with a foreigner can bring status, for Western women a relationship with an exotic Other can do the same. Here the arena of the public is different, though. On several occasions I watched Western women taking "selfies" with local men; some may just have been acquaintances or guides, but on a few occasions, the romantic involvement was obvious. In one instance, the procedure was elaborate: the woman kept repositioning the man in a way that would show some of the plaza and the mountain with the large Inca site as backdrop. She literally positioned him in line with Inca heritage and the past, the exotic Other staged and visually captured. We can see how the staging of indigenous culture for tourism, discussed earlier, plays out in personal relationships as well.

While both men and women can gain status among their peers from cross-cultural relationships, there is also strong criticism of these relationships. María, who frequently hosted foreign volunteers at her guesthouse, explained her disapproval to me:

> Volunteers, I warn them against the *bricheros,* especially the guys with the dreadlocks who sell jewelry and smoke marihuana. We had one girl here, I think from the US, and she went with one of those guys. But we knew him; he was bad, always going around with *gringas,* and he had three kids already! I told her that she couldn't bring him into the house. At first she was upset, but then she realized he was no good [...] Later she found another boyfriend who was much nicer.

The statement that the volunteer had found a "much nicer" boyfriend indicates that criticism was not directed at these cross-cultural relationships per se, but at the improper and exploitative behavior of certain men. This was illustrated on another occasion, when during one of my frequent trips to Cusco I sat next to María. Squished together on the back seat, we chatted and waited for the bus to fill up and leave. At the last minute, one of the street vendors jumped into a seat at the front. María turned to me with an expression of disgust and whispered "*Cochino* [dirty pig], this guy is a drug addict and takes advantage of women." I also heard other community members make critical comments about the *bricheros*, paralleling the situation in Jamaica, where local people tend to criticize men for living off foreign women and may even refer to them as prostitutes (Pruitt and LaFont 1995, 432–3). In Ollantaytambo I never heard *bricheros* called prostitutes; this term only came up in reference to women, as I will discuss. However, the men were clearly not viewed as positively as in Cusco, where conquering foreign women is often associated with virility and strength (Hill 2007, 444). My observations are more in line with those of Annelou Ypeij, who writes that, through their frequent contact with foreigners, the men often spend more time in bars, use their income to buy consumer goods, and dress more like Westerners. Their values and habits tend to shift, and they exhibit greater sexual promiscuity, which can lead to alienation from their own communities (2012, 27). Criticism was also racialized in that *bricheros* were sometimes referred to in derogative terms usually reserved for indigenous people. For example, a restaurant owner commented, "Why do the *gringas* always pick the ugliest Indian [*cholo*] they can find?"

Another interesting point is that the criticism I heard was predominantly directed at the men and not the foreign women. This contrasts with Susan Frohlick's findings from Costa Rica, where female travelers are often judged as sexually transgressive and deviant (2016). As discussed, female gender roles have become more flexible and permissive in recent years; nevertheless, a woman traveling on her own is still considered an anomaly and often met with a mix of curiosity, admiration, disdain, and pity. It is possible that since I was a woman living and traveling alone, people were careful in voicing their criticism to me; it may also be that some local people felt apologetic for the behavior of their own community members and wanted to warn me explicitly. It seems, then, that the main motive for men is material support while for women, the attraction of the exotic Other plays a major role. Both partners also seek a sexual experience and can gain status among their peers; however, the men in particular may be criticized by and alienated from other community members.

The concept of the *brichero* is clearly gendered and reflects ideologies of *machismo*. Many more local men seek relationships with foreigners than do local women (Babb 2012, 40), though the female version, the *brichera*, does exist. During my stay in Ollantaytambo, I heard of only two such cases. Adriana, a local woman introduced earlier, commented on the gender differences:

> I know many guys here who've had relationships with foreigners, and many have gone abroad. One went to France; he has a daughter there, but then he came back. My brother went to England.... One is in Australia; I think he is happy. That's what many guys here do, but many also come back. Women we don't do that so much. Frankly I have never felt attracted to a foreigner. Well, if it happens, good, but I'm not looking for it.

In terms of New Age themes, it is interesting to note that local women who seek relationships with foreigners do not change their appearance to the New Age Inca look that local men tend to cultivate (Bauer 2008, 613). However, the most prominent gender difference consists of the much harsher social judgments women experience. When I asked Naida about this, she responded that it was considered far more acceptable for local men to seek foreign women than vice versa. "They talk really badly about women who do that," she added. "It's like they prostitute themselves. Here you don't see it that much, but in Cusco many girls try to meet foreigners." On another occasion, I raised the topic with three young women, all craft vendors who had gathered to chat on a quiet afternoon. One woman seemed almost offended at the idea of a relationship with a foreigner and made it clear that she "would never, never do that." Another just commented that "people here talk a lot; you have to be careful." This perspective is in line with other findings on *bricheras* being considered akin to prostitutes (Hill 2007, 444; Meisch 1995). Also, I never saw public displays of affection between a Peruvian woman and a Western man, while local men conducted their relationships with foreigners quite openly. This reflects the fact that women are chastised far more for what are considered inappropriate relationships, while local men's pursuits are generally condoned or even encouraged by a culture of *machismo*.

While the Andean New Age trope of feeling is effectively mobilized in these cross-cultural relationships, what role do actual emotions play? Most research on sexual and romantic relationships in tourism emphasizes purpose and strategy, while feelings are often neglected (Tucker 2003, 138). Related to this, the emotional involvement and cost of these relationships may be underestimated. During my last meeting with Giulia, it was obvious that her relationship had taken an emotional toll on her. She looked tired and told me, "One moment he's super sweet, and then he is really mean. I feel like a stupid child. But whatever, I don't give a shit anymore; in two days I'm leaving for Bolivia." On another occasion a young American volunteer spoke to me about the heartbreak she had experienced with a Peruvian man. But it seems that most women enter these relationships with the expectation of a travel adventure rather than a long-term connection. Giulia had quickly moved on. When telling me about how the Peruvian had made her relax and feel more, Nina, the Dutch woman described earlier, had used the past tense: "I needed that." When I asked her if she was staying in touch with the man, she responded, "Sure, we are texting and Facebook, but well, I can't take him to Holland!" It was evident that, for her, the relationship was part of her travel experience and now finished.

Many writings highlight the exploitative nature of men in the context of sex and romance tourism, and in Peru, emotional sensitivity is strongly discouraged by a culture of *machismo*, which contributes to a gendered view and a disregard for men's emotional experiences. Yet the *brichero* in Mario Guevara Paredes's short story describes falling in love with a South African woman and ending up heartbroken and out of money. He puts the emotional challenges bluntly, saying that he knows "a lot of *bricheros* who aged prematurely because of such a rough life, and the *gringas* won't give a fucking buck for them now" (Guevara Paredes 1998, 101). To me, the emotional complexities for the men involved became evident in the comments of Joaquin, the bartender. He was around forty years old and originally from the coast. For a few weeks, he and I met regularly in a local café where I gave him English lessons in exchange for interview time. One of these afternoons, after we had finished with English past tenses, Joaquin told me about a recent relationship he had had with a French volunteer. She had left to carry on her trip around South America but later returned to spend another two weeks with him. Even though she kept sending emails, Joaquin had decided to cut off contact. "I guess her leaving affected me more than I thought," he told me. "Sometimes I wonder: did she just come back for sex tourism?" When I teased him lightly about the hardship of this, he looked upset and responded that he was now looking for something more permanent and did not want to feel used. I realized that my tactless response reflected some of my own gender bias that men are less emotionally vulnerable than women. Another case was Guillermo, a waiter of about thirty, who had lived abroad for a few years. He told me at length about an English woman he was in love with but who kept "just playing" with him, and his emotional pain was evident. On a different occasion, he also mentioned that he "kept prostituting himself." As mentioned earlier, I had not heard local men referred to as prostitutes, but the way Guillermo used the label for himself indicated a certain level of shame. Although a culture of *machismo* largely condones their liberal sexual behavior, men are not entirely free from emotional pain and inner conflicts of this behavior. In his discussion of masculinities and neoliberalism, Joe Hayns has documented a similar situation among Moroccan market vendors who engage in relationships with European women (2016). As these cases show, cross-cultural relationships are driven by more than strategic motives; they are also entangled in uncontrollable factors, such as emotions, and cultural biases may lead us to perceive these relationships in strongly gendered ways.

GENDER AND POWER

Latin America's colonial history is shaped by images of dominant White men and subjugated non-White women, two dimensions of inequality intersecting and compounding each other (Weismantel 2001, 155). Local *bricheros* and their predominantly White female partners reverse the historical pattern in terms of gender relations, but power inequalities still surface in different ways. Here it is useful to reconsider Nelson Graburn's analysis of travel as a secular ritual. As people leave home, cultural restrictions are lifted in the liminal space of the journey (Graburn 1989). While this approach has been criticized for its binary assumptions of home and away, it still highlights some prominent patterns of travel, which are negotiated in gendered ways. Since women arguably face greater restrictions in most societies, traveling can be a more empowering experience for them than for men. Israeli women traveling in India (Maoz 2008) and Western women in the Caribbean (Kempadoo 2004; Pruitt and LaFont 1995) both reported feeling liberated and empowered on their journeys, which sometimes included relationships with local men. However, women in search of sex and romance are also often judged as immoral and transgressive (Frohlick 2010, 2016).

In her study of cross-cultural relationships in Otavalo, Ecuador, Lynn Meisch writes that many Western women reported receiving a lot more attention from local men than they would at home (1995, 451). In a similar vein, Jacquie described the following experience from her guesthouse:

> My friend Claudio, he is twenty-four; I've known him since he was eighteen, and he's had plenty of women, like middle-aged women.... I've had women stay with me; I have noticed all these

> middle-aged women come in. One stayed for eight months. She would go on tours with these young guys and just, you know, women just getting all that attention they haven't gotten from men in their country ... just a whole other level of depth that these men help the women feel.

Jacquie's comment, as well as those I heard from other women, emphasized the surprise and pleasure of Western women receiving more male attention than they were used to. Another point worth noting is the factor of age, as in many of the relationships we find middle-aged women with significantly younger men. As quoted earlier, Jacquie spoke cheerfully about her "midlife adventure" and the joy of meeting "all these beautiful young men." For Western women, a relationship with a younger man is a reversal of cultural norms from home and arguably an empowering experience. Similar dynamics have been described for the Caribbean (Kempadoo 2004) and India (Maoz 2008).

Personally, I found that younger men would approach me fairly regularly with suggestions of having a drink or going for a hike. My strategy was to decline jokingly and mention that I was forty, an argument that did not appear to carry much weight. One man kept putting his arms around me declaring how he liked German women, while others just complimented me on looking younger and said they did not mind my age at all. Just as Western women are attracted to a simplified notion of the exotic, the appeal of the White Other may make age a secondary consideration for local men. Plus, if material support is a main motive, an older partner is likely to provide more support than a nineteen-year-old backpacker.

Returning to the concept of liminality, I think it worth noting that not just tourists experience this state. Annelou Ypeij writes that tour guides are "often associated with the figure of the *brichero* because in the eyes of their families, friends, and other local community members, they live between two worlds" (2012, 27). In Turkey, contact with foreigners also creates a liminal space for local men (Tucker 2003, 139–41):

> The liminal nature of both the women's and men's experiences in this tourist realm allows for and promotes a sense of romantic and sexual freedom that might be more restrained in their "home"

> contexts. He is in his new paradise where uncovered and "free" women are plentiful; she has arrived in an enchanting landscape where she is charmed by numerous attractive and attentive men.

This point constitutes a useful addition to Nelson Graburn's concept. The liminal effect for locals working in tourism is likely stronger the more gender roles differ between their own culture and those of visitors. Similarly, in the context of his work in Sumatra, Andrew Causey speaks about a "utopic space," referring to events and interactions that play out "in between reality and unrealizable desire." Here "the interactions between tourate and travelers indicate a common interest in exploring ways of being and acting in a space where their respective cultural rules and mores are suspended" (2003, 166–7).

However, foreigners have, in almost all cases, the advantage of greater wealth and mobility, a ticket back to their other lives, which causes a fundamental inequality in these relationships. In their analysis of Jamaican men's relationships with Western women, Deborah Pruitt and Suzanne LaFont also locate the power as mostly resting with the women (1995). In some cases, men may take advantage in more aggressive ways and sway the power balance in their favor, at least temporarily. The owner of a small guesthouse told me about one of his employees, a young local man:

> He spoke English really well, and I arranged for him to do some work as a guide. But mainly he just went out with the women; he never wanted to guide older people or couples. Always with the women, that was his thing. And sometimes he was drunk [...] He took advantage. There was this case when – I think it was an American woman – they spent a lot of time together when she was here, like boyfriend and girlfriend. Later she sent money to him; he was supposed to send crafts for her to sell in her country. But he never did. Hundreds of dollars, and he just kept it.

Clearly the young man had succeeded in his quest for money, but the American proceeded to denounce him and the guesthouse on online booking platforms, resulting in significant financial losses for the business. The greater online expertise many foreigners have provides another avenue through which to exercise power.

The effects of these relationships beyond the couple also need to be considered. In the context of a Turkish village, Hazel Tucker speaks of a "triangular set of relations ... between tourist women, local men and local women" (2003, 137). Again, Jacquie reports on this:

> I know that has hurt a lot of the women here because their partners start running around with the *gringas*. I had to ask some women to leave, some women staying long term and they got involved with some of my buddies whose wives are also my friends, and I had to tell them – "You know, you can't bring him here to my hostel. His wife is my friend!" [...] The women here know that their men are running around with *gringas*. But it also brings in money for their families, so they don't like it, but they deal with it. I know a number of women like that.

This comment points to the complex situations tourist-local relationships can create in the men's home lives, but it also confirms the financial importance of these relationships. Similarly, in the film *Cowboys in Paradise*, we can see how some local men on the island of Bali fund their families, including the construction of houses and shrines, through money from their relationships with foreign women (Virmani 2009). Another aspect is that local wives and partners often do not share the more liberal attitudes men may have acquired through their contact with foreigners, causing complications in their relationships (Ypeij 2012, 27).

It is important to consider that "dominance and power are not static but shifting and situational, constantly negotiated and contested, a process at once global and personal" (Pruitt and LaFont 1995, 437). We can see these relationships as a field in which this negotiation plays out; while usually the Western women have dominance in terms of financial means, men also have the opportunity to gain power, at least temporarily. The length of the relationship is another important factor. If a woman decides to stay long term, this affects the power dynamics in different ways. Jacquie had been in a relationship with a local man for several years. She told me how the two had started running a hostel together:

> It was really challenging; it was a brand new relationship and a business partnership as well. And so I provided all the money … that's what happens here, and so I took that on. And he and I started going to Cusco to buy everything.[...] It pains me ethically to see *gringos* coming in and doing business that doesn't involve local people in any way. And at the same time, it has been very challenging doing it with a local partner; it's been power struggles. I mean, I brought all this money, so in my culture this would make me more invested. Not in this culture, because this is his godfather's house, his town, his street, his culture. So for him, he is more invested, you know? So we have that struggle.

Even though Jacquie brings economic, social, and cultural capital to apply successfully in the tourism business, this does not necessarily translate into a stable power position. Furthermore, in long-term relationships, the liminal state fades and both partners are faced with more traditional gender expectations. In Turkey, local cultural norms mean much greater restriction for women in terms of movement and interaction, as well as fulfilling social obligations for their male partners' extended family – an unfamiliar and often very challenging role that differs strongly from the freedom these women had experienced as tourists (Tucker 2003, 147). To a lesser degree, I heard similar comments in Ollantaytambo. Jacquie explained:

> I guess for me the most challenging part was not just opening a business and dealing with tourists, but also this town. It is so small, and my partner is a home boy, so all of a sudden there is this new community, new culture. We were living there in one room; it was the only way to do it then. So yeah, I just never had any time to myself; for three years pretty much 24/7 I was playing hostess to tourists or locals.… That's why right now I have my own place.

On the one hand, Jacquie now seems quite well integrated into the community. I often saw local people visiting her house, and she and her partner have been invited to hold an important role in a major upcoming community celebration. On the other hand, Jacquie's position remains ambiguous. Before meeting her in person, I had heard local people talk about her. Three of them referred to her as a witch (*bruja*). One man told me to be careful that she did not get any of my hair or she would use it in one of her rituals. The comment was made in jest, but the man still seemed to consider it a possibility. While these accusations may have to do with the fact that she and her partner run spiritual ceremonies out of their hostel, I think it also reflects that many community members do not trust the foreigner completely, resulting in a more precarious situation for Jacquie. Once women become more invested emotionally and financially, they lose a lot of the power and independence they held as tourists, while at the same time, they may never achieve full community support either. As discussed earlier, even well-integrated foreigners can become targets of aggression, and women are more vulnerable if this becomes physically violent.

As we have seen, spiritual and romance tourism intersect in complex ways. Both involve problematic aspects of commodification but can also allow local people to challenge unequal power relations, at least temporarily. As Michel Foucault reminds us, resistance is an inherent part of power relations, but is also often short-lived (1978, 95–6). It is doubtful that mystical tourism can help address structural problems of inequality, but it has the potential to contribute to greater interest in cultural differences (Hill 2008, 273). Mystical tourism clearly perpetuates essentializing and racializing stereotypes, and urban *mestizos* and foreigners commodify indigenous cultural knowledge and reap most of the benefits. However, in the often alienating day-to-day interactions with tourists, New Age discourse can function as a form of "veiled resistance" (Boissevain 1995, 14–15; Maoz 2006, 224) and offer a more empowering alternative to the common narrative that positions local people largely as helpless victims. It is useful here to consider the concept of self-commodification, "a set of beliefs and practices in which an individual chooses to construct a marketable identity product while striving to avoid alienating him- or herself." The process is often contradictory but emphasizes the personal agency in people's response to larger capitalist structural forces (Bunten 2008, 381). Whereas in large foreign run centers, indigenous spiritual and cultural capital are often appropriated and commodified by outsiders, in most instances I witnessed, local people exercised a significant amount of control over how to present themselves and their culture, and they knew how to draw on New Age discourse for their benefit. One early evening when I was chatting with José, he pointed to the Inca rocks in the foundations of the houses. "When the sun goes down, the whole street turns golden, have you noticed? Everyone is looking for the golden city, but ha – they can't see. And here it is right in front of them: the stones turn to gold!" New Age tourism has become big business, yet José's message – that we do not need to go far but just open our eyes to what is right in front of us – captures an essential point that most spiritual tourists miss.

Bricheros effectively respond to tourist imaginaries by mobilizing local mythologies and New Age themes. While the concept of self-commodification (Bunten 2008) is useful in highlighting the personal agency of the *bricheros*, it sidelines the larger structural power inequalities involved. Research in the Caribbean has shown that the dynamics and beliefs expressed in these cross-cultural relationships often replicate the power and wealth differential between the partners' home countries (Kempadoo 2004; Pruit and LaFont 1995). Similarly, *bricheros* reflect the contradictions of modernity in Peru's neoliberal economic climate, and in order to access some of the wealth of foreigners, locals often have to resort to presenting themselves as the "primitive mystic" (Hill 2007, 446). Kamala Kempadoo convincingly argues that the distinction between sex tourism and romance tourism is embedded in Western hegemonic gender norms that position men as sexually aggressive and approaching love and sex as separate, whereas women are considered passive and more concerned with love and intimacy than sex. Thus, framing their relationships as romance rather than just sex aligns Western women with their gender norms and allows them to differentiate their actions from male sex tourism (Kempadoo 2004, 128).

Further contradictions emerge in terms of masculine identities. Men's responses to neoliberalism frequently include the struggle for, but also the disillusion with, the role of provider (Cornwall 2016). Earlier, I described how local men can act aggressively when feeling threatened in their capacity to fill this role and how tourism affords women new work opportunities that may clash with traditional gendered expectations. While *bricheros* living off foreign women appear to reject their role as provider, channeling resources into their local families also allows them to comply with these gendered expectations in a novel way.

Last, it is also worth noting that Western women and Peruvian men have a shared experience of oppression, the former based on gender and the latter based on racialized identities and colonial history. In that regard, they meet on a somewhat more equal footing than we typically find in the context of sex tourism between Western men and local women, where inequalities based on gender and nationality are compounded. Furthermore, it has been argued that power is not inherently linked to sex and the male domain but rather to specific behaviors and ways of being (e.g., Cornwall 1997, 11). My findings are in line with those of others (Kempadoo 2004; Herold et al. 2001) who state that both women and men can exercise dominance and power, if one gender is able to access more material resources than the other. Just as colonial love stories usually end with the European returning home (Pratt 1992, 100), today's relationships between local men and Western women afford women the power to do the same. Giulia was clearly traveling on a shoestring budget, but when the relationship with the local man did not meet her needs anymore, she had the means to move on.

7

Conclusion

As the COVID-19 pandemic brought travel to a halt, the World Tourism Organization estimated a 60 to 80 percent drop in all international travel for 2020 worldwide (UNWTO 2020a). Travel played a major role in spreading COVID-19, so in many places people have become worried about the arrival of foreigners. Yet, UN Secretary-General António Guterres has stated that "tourism can be a platform for overcoming the pandemic. By bringing people together, tourism can promote solidarity and trust" (UNWTO 2020b). Despite the current challenges and uncertainties, tourism is likely to resume its growth, and thus it becomes even more important to understand its complex effects and to reflect on our own roles in it, both as anthropologists and as tourists. In this book, I have examined how inequalities in terms of gender and power can be exacerbated, negotiated, and also challenged in the context of tourism. These dynamics are often specific to the locations and cultural contexts in which they unfold, and care must be taken when generalizing findings. However, the in-depth study of one location can also reveal aspects of tourism impacts that apply more broadly. Like many of the tourists I studied, as a White Western anthropologist, I, too, benefitted from global structural inequalities, and my background may have motivated some Peruvians to participate in my research. Although I disclosed my purpose and offered different forms of compensation, it is important to acknowledge that this work is built partially on the colonial history shared by both tourism and the discipline of anthropology.

I conclude by briefly revisiting a few key points and then addressing recent changes and recommendations for both local people and tourists. Tourists, visitors, and brokers experience each other through a "mutual gaze" (Maoz 2006) and complex tourist imaginaries (Salazar 2012). Peru's two prominent industries of mining and tourism are both dominated by foreign and national elites and have led to greater disparities in wealth (Steel 2013, 245), yet tourism discourse reveals different ways in which visitors legitimize their presence and distance themselves from these inequalities. Two years after architect Graham Hannegan pointed to the growing disparities visible in house and roof construction in Ollantaytambo, a new tourist video about the village published on Facebook showed a number of expanded and new buildings with different types of roofs, attesting to the ongoing changes. Most of the town's adobe walls have been painted yellow, as shown in this book's cover image. Naida told me that local people are moving away in increasingly large numbers, as the majority of houses in the old part of Ollantaytambo are converted into tourist restaurants and accommodations. In the city of Cusco, rapidly rising land and property prices have already pushed local people out of the historic center (Steel 2013, 243), and this pattern now seems to be replicated in smaller communities. María and her family have finished building their new home outside of town. Their property in the country is larger, and she frequently posts photos of her beautiful garden on Facebook. One of them shows her son smiling and holding up a big basket of peaches. She and her family seem to be doing well, but what do these trends mean for the community long term?

My research indicates that small, flexible business opportunities have been particularly beneficial for women as these provide access to paid work, which is generally valued more, and can be combined with existing gendered tasks, particularly childcare. Adriana gave birth to a baby boy three years ago. The child's father is absent, and she continues to live with her parents and helps with managing their small guesthouse. "I'm happy with my little prince," she tells me. "We are doing well. I just hope there will be more tourists; please tell your friends to come visit." It seems that through increased work opportunities in public, women's presence there has become more acceptable. Also, as formerly unpaid domestic chores like cooking and cleaning become paid labor, their value increases. Many young men spoke with pride about their work as cooks in tourist establishments, and there are signs that they are also picking up more of these tasks at home. The presence of foreigners demonstrating less rigid gender relations, and especially young women traveling independently, has also had an effect on local perceptions. However, despite women's greater involvement in work outside the home, inequalities remain. Men often have access to more prestigious and better paying work than women, and most guide jobs in Ollantaytambo are held by men. For example, Diego's work as a tour guide provides most of his family's income. In addition, men may feel threatened in their position of power in the home, sometimes leading to a violent backlash against women. Also, local family businesses like Adriana's may face more difficulties in the future. Already during my stay, people complained about lower numbers of tourists, which surprised me as statistics showed a continuous increase at the time. The reason for this is likely a shift of tourism infrastructure and services. For many years, the region attracted the more adventurous, independent travelers, and high-end accommodation was exclusive to urban centers, but this trend has been changing. More visitors now visit on prearranged tours, which often organize meals in affiliated restaurants and leave little time for tourists to explore and use the smaller local businesses. During my stay, several large, new restaurants opened along the main road of the Sacred Valley where tour buses stop. If we consider women's empowerment as a more collective than individual process that opens a greater range of choices for women (Apffel-Marglin and Sanchez 2002; Mosedale 2005, 252), then the growing competition from larger businesses is problematic since it may once again reduce women's choices of flexible work.

While Ollantaytambo is changing to provide more tourism infrastructure, it is simultaneously attempting to meet visitors' expectations of authenticity. Aspects of indigenous culture become commodified, as seen, for example, in the decorative displays on restaurant walls. More agency can be seen by the women and children from nearby indigenous communities who pose for photos. With their community-managed rotation system, they seem to have found an effective way to regulate tourism work in a more egalitarian fashion, which could provide a good model for other communities to follow.

Tourism development and growing material inequalities also result in aggression between community members. Joaquin left Ollantaytambo and his work at the bar and now runs an import business in Lima; he told me that dealing with the ongoing conflicts had become too difficult. While resentment toward tourists was usually articulated indirectly through storytelling and humor, aggression between community members and toward foreign residents was often more open. *Mestizo* gender ideologies still define men as providers, so they are often the ones who feel the competition from foreigners most acutely, which partly accounts for the expressions of open conflict. It is important to recognize the relational dimensions of gender issues and understand how patriarchal structures can limit both women and men (Cornwall 2014, 136). Social tensions are also revealed by ongoing cases of black magic. During a phone conversation two years ago, Naida told me that they had had fewer guests, which she attributed to witchcraft performed by envious neighbors. Based on the notion of "limited good" (Foster 1965), making sure a neighbor does not prosper too much is considered to increase one's own chances of success. With the current pandemic and tourism halted, some of these conflicts may ease off, while resentment toward tourists may take on a new dimension involving fear of infection.

Other aspects marketed for tourism are spirituality and romance. On the one hand, this type of tourism constitutes a selling out to foreign demands and involves the appropriation of indigenous cultural capital by *mestizos* and foreigners. On the other hand, the tropes of New Age discourse can also serve as a strategy of resistance that allows local people to position themselves in a more knowledgeable and powerful role vis-à-vis foreigners. Power relations also come into play in the personal arena of romance. While the relationships between Western women and local men tend to be more balanced than instances of sex tourism elsewhere, they still replicate many of the global inequalities of the partners' nationalities. Facilitated by global economic disparities, Giulia has continued working summers in Germany and spending winters in warmer locales. While last year her Facebook photos showed her smiling on a sunny beach in Costa Rica, this year the pandemic forced her to stay in Germany. American resident Jacquie has left her partner and now lives in Cusco. Hers was the longest cross-cultural relationship I encountered during my fieldwork, but it appears that ultimately the differences were too difficult to bridge.

In many ways, tourism encounters recreate colonial relationships as the often wealthier White visitors are served by local people, and aspects of local culture and belief systems become commodified and offered for consumption. However, my findings indicate that tourist relations are more complex and dynamic for a conclusion that simple. Rather than deciding whether certain acts should be considered forms of resistance or reflections of historical and neocolonial power inequalities, one must consider how these aspects intersect and coexist (Hollinshead 2004). While colonial history and contemporary global inequalities frame encounters between locals and visitors, it is also evident that people actively renegotiate and challenge these inequalities in different ways.

RECOMMENDATIONS AND FINAL THOUGHTS

As of September 2020, Peru is under a second lockdown due to the pandemic. Diego posted photos on Facebook showing the usually busy streets of Ollantaytambo deserted and the craft market boarded up. He now works in construction in nearby Urubamba in order to support his family. In a recent conversation, María told me that both she and her husband lost their work in tourism and have cleared an additional piece of land in order to grow potatoes and other crops. She described how the large parking lot for tour buses has been turned into an outdoor market that allows for social distancing. Many tourism-related businesses in town have closed for good. A young couple, whom I remember as working tirelessly in their small restaurant, recently sold their entire inventory out on the street. Naida and Víctor have locked up their guesthouse and moved in with Naida's elderly parents in the country; they hope to reopen their business as soon as restrictions are lifted. More than half of the people I spoke with in Ollantaytambo during my research reported that some family members still worked in agriculture, which is currently providing a crucial economic fallback system. The COVID-19 crisis has caused the mass migration of suddenly unemployed urban workers back to their rural home communities, underscoring the importance of this safety net.

Another factor that impacts the long-term sustainability of tourism in the Andes, and people's livelihoods more broadly, is climate change. Resulting erosion and soil degradation affect farming and food security (Zoomers 2008, 974); this environmental damage is exacerbated by changing weather patterns and melting glaciers, which have functioned as water reservoirs in the past. Conflicts about water management, already a problem in many parts of the Andes, are expected to increase significantly in the coming years (Murtinho et al. 2013). When local people spoke to me about work opportunities outside of tourism, these mostly fell into the two categories of "working in the fields" locally or moving away to the cities or lowlands for other types of work, notably mining and factory or plantation work. The scarcity of work options in the highlands has led to a strong economic dependence on tourism. Considering the precarious nature of tourism work, it is advisable to combine tourism development with other economic strategies.

The government continues to support tourism strongly as an avenue for economic growth (PROMPERÙ 2020). In hopes of boosting travel after the pandemic, it declared that entry to major archaeological sites like Machu Picchu, Pisac, and Ollantaytambo is free of charge from July to December 2020. A couple of years ago, the so-called Rainbow Mountain, or Vinicunca, with its striking colorful patterns caused by mineral deposits, became a major tourist attraction south of Cusco. This development close to Apu Ausangate, southern Peru's most sacred peak, has saved the area from more destructive mining, at least for now (Lombrana 2018). In addition, the Amazon region has been increasingly promoted for tourism development (Steel 2013, 246), so we can expect to see similar developments there as in the popular highland region. However, few locations have the characteristics to attract large numbers of tourists consistently. In that regard, Ollantaytambo stands out. Its proximity to Cusco and Machu Picchu, unique Inca architecture, and beautiful natural surroundings are factors that consistently make it attractive to visitors. On the other hand, these same factors may also make it difficult to limit development, should the community so choose. As mentioned earlier, Ollantaytambo does not allow the sale of land to foreigners, but that is not the case for the

Sacred Valley in general, which has seen a large inflow of foreigners and new construction. In Ecuador, the growth of tourism paved the way for international residents to move in, and, attracted to rural regions and low land prices, foreign residents contribute to the further marginalization of rural people (Gascón 2016). Around the town of Pisac, at the other end of the Sacred Valley, foreigners have already displaced local farmers from valuable agricultural land. Both tourists and locals have identified garbage and waste management as major problems, and clear regulations are needed to protect local people and the environment.

Studies in other parts of the Andes have shown that, as people become wealthier, they often turn away from long-standing cultural practices (Zoomers 2008, 980). However, there are also counterexamples. In the Ecuadorian Andes, the money made from the international craft trade has allowed people to stay in rural communities and continue their cultural traditions rather than migrate to urban centers (Colloredo-Mansfeld 1999). This effect is evident to some extent in Ollantaytambo as well. Alberto told me, "We all used to want to go to Cusco or Lima. But now there's more work here, so we are happy to stay." In addition, tourist interest brings some validation to local culture, which is particularly relevant in a society where rural and indigenous populations have long been marginalized. Thus, the notion that increased tourism always leads to cultural loss cannot be substantiated.

It is also important to consider that "tourism discourse which promotes the preservation of the 'traditional' for tourist experience is itself based on a colonial desire to fix the identity of the other in order that it remains (or perhaps in actuality becomes) distinct from tourist identity" (Hall and Tucker 2004, 17). Rather than focusing on the preservation of often narrowly defined cultural traditions, discussions about sustainability need to include new cultural developments (Tucker 2003). This consideration is also relevant to the concept of authenticity, which still emerged as a strong concern for tourists searching for "the timeless Andean village." Erik Cohen's concept of "emergent authenticity" (1988) helps us move beyond static definitions and acknowledges that cultural forms are always changing. Another important factor is control. Authenticity can also be defined as "conditions in which people have significant control over their affairs, to the extent that they are able to play an active role in determining how changes occur in their social setting" (Chambers 2010, 101). As studies in other parts of the Andes have shown, a concentration of power in the hands of a few community members is detrimental, and, beginning in the early stages of tourism development, as many residents as possible should be involved (Mitchell and Eagles 2001, 26). In Ollantaytambo, many people felt a lack of control over the rapid developments and did not see their concerns heard and addressed by the municipal government. This imbalance in influence and control is gendered as well. It has been shown that women benefit far more from tourism development if they can actively take part in the design of projects (Ferguson 2010a, 20). In Ollantaytambo, women's participation in tourism is certainly significant, and the association of craft vendors, who are almost all female, manages many of its members' own affairs.

Community unity, defined as "collective support for the local tourism sector and community cohesiveness," has been identified as one of the crucial elements of tourism success (Mitchell and Eagles 2001, 6). On Taquile Island in Lake Titicaca, almost all community members gain some of their income from tourism, either by providing accommodation or by selling handicrafts (Mitchell and Eagles 2001, 18). During my time in Ollantaytambo, most businesses were locally owned, but involvement in tourism was not as inclusive as on Taquile. Some obstacles are skills based, such as a lack of computer and English skills needed to access and manage different types of tourism work successfully. Business competition is increasing because of larger businesses and outsiders who bring more expertise. Benefits could also be distributed better through investment in community services. Rosa stated, "Tourism is growing so much; we need a good place right here to study and prepare for work. And we need to invest in the children. They don't even have a recreational center; they are just out on the street." Her comment clearly speaks to the concerns of balancing individual with communal benefits in the context of growing tourism development. Ollantaytambo's Inca foundations have stood strong for more than five centuries, used to support the town's colonial Spanish constructions and now tourist restaurants and hotels. Likewise, "the living Inca town" has preserved many of its cultural traditions. What might be most needed now is the continued practice of *ayni*, the Andean tradition of reciprocity, to balance individual gains and growing inequalities with solidarity and cooperation.

Last, tourists also have an important part to play. Stefanie, the German backpacker, expressed her struggle with the tourist role several times, commenting that she was trying to be a "good traveler" but was wondering how best to do that. Other tourists questioned if it was ethical to travel at all. Conversely, the majority of local people emphasized that they welcomed tourism and hoped for more visitors. The most common complaint I heard was about a lack of respect, which locals felt most strongly from coastal Peruvians and other Latin Americans. This disrespect was experienced through disparaging and impolite comments and through certain behaviors and styles of dress worn by women. Tourists can feel stressed by travel, and at times they are taken advantage of, yet these minor exploitations happen in the context of broader global inequalities and Western privilege. As this study has shown, power differentials shape tourist encounters in many ways, while at the same time, tourist imaginaries and discourse tend to minimize these disparities. This raises questions about if and how tourism can be conducted more ethically. While the underlying inequalities cannot be changed easily, tourists are able to make a difference through greater awareness and consideration of how they interact with other people and places. Many locals expressed appreciation for what may seem like small gestures. Daniela, the hotel manager, commented, "I really like that many tourists stop and talk to the children." Alberto told me, "It's good when visitors show an interest, when they ask about our culture and how we live. And I like learning about their country." The current pandemic crisis challenges us to reimagine tourism in more beneficial ways. In today's world, where encounters with those different from us do not just happen on extended journeys, it becomes ever more important to understand the specific ways in which we can find connection beyond our differences. While I have been critical of many aspects of tourism, I also still believe that travel, like anthropology, has immense potential to challenge assumptions, teach new perspectives, and help us recognize our basic commonalities as human beings.

References

Albers, Patricia C., and William R. James. 1988. "Travel Photography: A Methodological Approach." *Annals of Tourism Research* 15:134–58. https://doi.org/10.1016/0160-7383(88)90076-X

Allen, Catherine. 1988. *The Hold Life Has: Coca and Cultural Identity in an Andean Community.* Washington: Smithsonian Institution Press.

Apffel-Marglin, Frédérique, and Loyda Sanchez. 2002. "Developmentalist Feminism and Neocolonialism in Andean Communities." In *Feminist Post-Development Thought: Rethinking Modernity, Post-Colonialism and Representation*, edited by Kriemild Saunders, 159–79. London: Zed Books.

Appadurai, Arjun. 1990. "Disjuncture and Difference in the Global Cultural Economy." *Theory, Culture & Society* 7:295–310. https://doi.org/10.1177/026327690007002017

———. 1996. *Modernity at Large: Cultural Dimensions of Globalization.* Minneapolis: University of Minnesota Press.

Arellano, Alexandra. 2004. "Bodies, Spirits, and Incas: Performing Machu Picchu." In *Tourism Mobilities: Places to Play, Places in Play*, edited by Mimi Sheller and John Urry, 67–77. London: Routledge.

Ascione, Elisa (in press, fall 2020). "Teaching and Learning Food and Sustainability in Italy: Betwixt and Beyond Touristic Consumption." In *Study Abroad and the Quest for an Anti-Tourism Experience*, edited by Bodinger de Uriarte and Michael A. Di Giovine, chapter 9. Lanham: Lexington Books.

Aslanbeigui, Nahid, Guy Oakes, and Nancy Uddin. 2010. "Assessing Microcredit in Bangladesh: A Critique of the Concept of Empowerment." *Review of Political Economy* 22(2):181–204. https://doi.org/10.1080/09538251003665446

Babb, Florence E. 1989. *Between Field and Cooking Pot: The Political Economy of Marketwomen in Peru*. Austin: University of Texas Press.
———. 2011. *The Tourism Encounter: Fashioning Latin American Nations and Histories*. Stanford: Stanford University Press.
———. 2012. "Theorizing Gender, Race, and Cultural Tourism in Latin America: A View from Peru and Mexico." *Latin American Perspectives* 39(6):36–50. https://doi.org/10.1177/0094582X12454560
Bærenholdt, Jorgen O., Michael Haldrup, Jonas Larsen, and John Urry. 2004. *Performing Tourist Places*. Aldershot: Ashgate Publishing.
Bailey, Frederick G. 1971. *Gifts and Poison: The Politics of Reputation*. New York: Schocken Books.
Banks, Marcus. 2005. "Visual Anthropology: Image, Object, and Interpretation." In *Image-based Research: A Sourcebook for Qualitative Researchers*, edited by Jon Prosser, 6–19. Philadelphia: Falmer Press.
Barrig, Maruja. 2006. "What is Justice? Indigenous Women in Andean Development Projects." In *Women and Gender Equity in Development Theory and Practice: Institutions, Resources, Mobilization*, edited by Jane S. Jaquette and Gale Summerfield, 107–33. Durham: Duke University Press.
Bauer, Irmgard. 2008. "'They Don't Just Come for Machu Picchu': Locals' Views of Tourist-Local Sexual Relationships in Cuzco, Peru." *Culture, Health & Sexuality* 10(6):611–24. https://doi.org/10.1080/13691050802155376
Beh, Adam, Brett L. Bruyere, and Sam Lolosoli. 2013. "Legitimizing Local Perspectives in Conservation through Community-based Research: A Photovoice Study in Samburu, Kenya." *Society and Natural Resources* 26(12):1390–406. https://doi.org/10.1080/08941920.2013.805858
Behar, Ruth. 1996. *The Vulnerable Observer: Anthropology that Breaks your Heart*. Boston: Beacon Press.
Benería, Lourdes. 2003. *Gender, Development, and Globalization: Economics as if all People Mattered*. London: Routledge.
Binder, Jana. 2004. "The Whole Point of Backpacking: Anthropological Perspectives on the Characteristics of Backpacking." In *The Global Nomad: Backpacker Travel in Theory and Practice*, edited by Greg Richards and Julie Wilson, 92–108. Clevedon: Channel View Publications
Boissevain, Jeremy. 1995. *Coping with Tourists: European Reaction to Mass Tourism*. Oxford: Berghan Books.
Bolin, Inge. 1998. *Rituals of Respect: The Secret of Survival in the High Peruvian Andes*. Austin: University of Texas Press.
———. 2006. *Growing Up in a Culture of Respect: Child Rearing in Highland Peru*. Austin: University of Texas Press.
Bopp, Judie, Michael Bopp, Lee Brown, and Phil Lane Jr. 2004. *The Sacred Tree: Reflections on Native American Spirituality*. 4th ed. Lethbridge: Four Worlds International Institute.

Boudreault-Fournier, Alexandrine. 2012. "Ethnographic Mise-en-Scene: On Locating the Anthropologist's Creative Agency." *Anthropology News*. http://dev.aaanet.org/news/index.php/2012/02/13/ ethnographic -mise-en-scene/

Bourdieu, Pierre. 1984. *Distinction: A Social Critique of the Judgment of Taste*. Cambridge: Harvard University Press.

Bourque, Susan C., and Kay Barbara Warren. 1981. *Women of the Andes: Patriarchy and Social Change in two Peruvian Towns*. Ann Arbor: University of Michigan Press.

Bray, Zoe. 2015. "Anthropology with a Paintbrush: Naturalist-Realist Painting as 'Thick Description.'" *Visual Anthropology Review* 31(2):119–33. https://doi.org/10.1111/var.12076

Brulotte, Ronda L. 2017. "Archaeological Replica Vendors and an Alternative History of Mexican Heritage Site: The Case of Monte Albán." In *World Heritage Sites and Tourism: Global and Local Relations*, edited by Laurent Bourdeau, Maria Gravari-Barbas, and Mike Robinson, 56–66. London: Routledge.

Bruner, Edward M. 1991. "Transformation of Self in Tourism." *Annals of Tourism Research* 18(2):238–50. https://doi.org/10.1016/0160 -7383(91)90007-X

———. 2005. *Culture on Tour: Ethnographies of Travel*. Chicago: University of Chicago Press.

Bunten, Alexis C. 2008. "Sharing Culture or Selling Out?: Developing the Commodified Persona in the Heritage Industry." *American Ethnologist* 35(3):380–95. https://doi.org/10.1111/j.1548-1425.2008.00041.x

Butler, Judith. 1990 *Gender Trouble: Feminism and the Subversion of Identity*. London: Routledge.

Cahyanto, Ignatius, Lori Pennington-Gray, and Brijesh Thapa. 2013. "Tourist-Resident Interfaces: Using Reflexive Photography to Develop Responsible Rural Tourism in Indonesia." *Journal of Sustainable Tourism* 21(5):732–49. https://doi.org/10.1080/09669582.2012.709860

Cajete, Gregory. 1994. *Look to the Mountain: An Ecology of Indigenous Education*. Durango, CO: Kivaki Press.

Canessa, Andrew, ed. 2005. "The Indian Within, the Indian Without: Citizenship, Race, and Sex in a Bolivian Hamlet." In *Natives Making Nation: Gender, Indigeneity, and the State in the Andes*, 130–55. Tucson: University of Arizona Press.

Castañeda, Quetzil E. 1997. *In the Museum of Maya Culture: Touring Chichén Itzá*. Minneapolis: University of Minnesota Press.

Causey, Andrew. 2003. *Hard Bargaining in Sumatra: Western Travelers and Toba Bataks in the Marketplace of Souvenirs*. Honolulu: University of Hawai'i Press.

———. 2017. *Drawn to See: Drawings as an Ethnographic Method*. Toronto: University of Toronto Press.

Chambers, Erve. 2010. *Native Tours: The Anthropology of Travel and Tourism*. 2nd ed. Long Grove: Waveland Press.
Cheong, So-Min, and Marc L. Miller. 2004. "Power Dynamics in Tourism: A Foucauldian Approach." In *Tourists and Tourism: A Reader*, edited by Sharon Bohn Gmelch, 239–52. Long Grove: Waveland Press.
Cohen, Erik. 1972. "Toward a Sociology of International Tourism." *Social Research* 39(1):164–82. https://www.jstor.org/stable/40970087
———. 1988. "Authenticity and Commoditization in Tourism." *Annals of Tourism Research* 15(3):371–86. https://doi.org/10.1016/0160-7383(88)90028-X
———. 1995. "Contemporary Tourism – Trends and Challenges: Sustainable Authenticity or Contrived Postmodernity?" In *Change in Tourism: People, Places, Processes*, edited by Richard Butler and Douglas Pearce, 12–29. London: Routledge.
Collier, John Jr., and Malcom Collier. 1986. *Visual Anthropology: Photography as a Research Method*. Albuquerque: University of New Mexico Press.
Colloredo-Mansfeld, Rudi. 1993 "The Value of Sketching in Field Research." *Anthropology UCLA* 20:89–104.
———. 1998. "'Dirty Indians', Radical *Indígenas*, and the Political Economy of Social Difference in Modern Ecuador." *The Bulletin of Latin American Research* 17(2):185–205. https://doi.org/10.1016/S0261-3050(97)00087-9
———. 1999. *The Native Leisure Class: Consumption and Cultural Creativity in the Andes*. Chicago: University of Chicago Press.
Cornwall, Andrea. 1997. "Men, Masculinity, and 'Gender in Development.'" *Gender & Development* 5(2):8–13. https://doi.org/10.1080/741922358
———. 2014. "Taking off International Development's Straightjacket of Gender." *The Brown Journal of World Affairs* 21(1):127–39. http://bjwa.brown.edu/21-1/taking-off-international-developments-straightjacket-of-gender/
———. 2016. "Introduction: Masculinities under Neoliberalism." In *Masculinities under Neoliberalism*, edited by Andrea Cornwall, Frank G. Karioris, and Nancy Lindisfarne, 1–28. London: Zed Books.
Costa, Joana, and Elydia Silva. 2010. "Gender Inequalities and Poverty: A Simulation of the Likely Impacts of Reducing Labour Market Inequalities on Poverty Incidence in Latin America." In *The International Handbook of Gender and Poverty: Concepts, Research, Policy*, edited by Sylvia Chant, 490–94. Cheltenham: Edward Elgar.
Cox Hall, Amy. 2017. "Machu Picchu: An Andean Utopia for the Twenty-First Century?" In *World Heritage Sites and Tourism: Global and Local Relations*, edited by Laurent Bourdeau, Maria Gravari-Barbas, and Mike Robinson, 37–44. London: Routledge.
Crick, Malcolm. 1989. "Representations of International Tourism in Social Sciences: Sun, Sex, Sights, Savings, and Servility." *Annual Review of*

Anthropology 18:307–44. https://doi.org/10.1146/annurev.an.18.100189.001515

Dann, Graham. 1996. "The People of Tourist Brochures." In *The Tourist Image: Myths and Myth Making in Tourism*, edited by Tom Selwyn, 61–81. Chichester: John Wiley & Sons.

Davalos, Karen Mary. 2008. "*Sin Vergüenza*: Chicana Feminist Theorizing." *Feminist Studies* 34(1/2):151–71. https://www.jstor.org/stable/20459186

Davis, Wade. 1996. *One River: Explorations and Discoveries in the Amazon Rain Forest*. New York: Simon & Schuster.

de la Cadena, Marisol. 1995. "'Women Are More Indian': Ethnicity and Gender in a Community near Cuzco." In *Ethnicity, Markets, and Migration in the Andes: At the Crossroads of History and Anthropology*, edited by Brooke Larson, Olivia Harris, and Enrique Tandeter, 329–48. Durham: Duke University Press.

———. 2003. *Indigenous Mestizos: The Politics of Race and Culture in Cuzco, Peru, 1919–1991*. Durham: Duke University Press.

del Casino, Vincent J. Jr. 2009. *Social Geography: A Critical Introduction*. Chichester: Wiley-Blackwell.

Desmond, Jane. 1999. *Staging Tourism: Bodies on Display from Waikiki to Sea World*. Chicago: University of Chicago Press.

Di Giovine, Michael A. 2009. *The Heritage-Scape: UNESCO, World Heritage, and Tourism*. Lanham: Lexington Books.

Downe, Pamela J. 1999. "Questions for the Questioner: Feminist Methodology and Anthropological Research." In *Exploring the Social World: Social Research in Action*, edited by Dawn Currie, David Hay, Brian MacLean, 129–43. Vancouver: Collective Press.

Dubois, Laurent. 1995."'Man's Darkest Hours': Maleness, Travel, and Anthropology." In *Women Writing Culture*, edited by Ruth Behar and Deborah A. Gordon, 306–21. Berkeley: University of California Press.

Duran, Jane. 2001. *Worlds of Knowing: Global Feminist Epistemologies*. New York: Routledge.

Eber, Christine. 1995. *Women and Alcohol in a Highland Maya Town: Water of Hope, Water of Sorrow*. Austin: University of Texas Press.

Edwards, Elizabeth. 1996. "Postcards – Greetings from Another World." In *The Tourist Image: Myths and Myth Making in Tourism*, edited by Tom Selwyn, 197–221. Chichester: John Wiley & Sons Ltd.

———. 1997. "Beyond the Boundary: A Consideration of the Expressive in Photography and Anthropology." In *Rethinking Visual Anthropology*, edited by Marcus Banks and Howard Morphy, 53–80. New Haven: Yale University Press.

Elsrud, Torun. 2006. "Gender Creation in Travelling, or the Art of Transforming an Adventuress." In *Tourism Consumption and Transformation: Narratives of Place and Self*, edited by Kevin Meethan, Alison Anderson, Steve Miles, 178–95. Wallingford: CABI.

Enloe, Cynthia. 2000. *Bananas, Beaches and Bases: Making Feminist Sense of International Politics*. Berkeley: University of California Press.

Eriksen, Thomas Hylland. 2015. "What's Wrong with the Global North and the Global South?" https://www.hyllanderiksen.net/blog/2018/12/13/whats-wrong-with-the-global-north-and-the-global-south/

Errington, Frederick, and Deborah Gewertz. 1989. "Tourism and Anthropology in a Post-Modern World." *Oceania* 60(1):37–54. https://doi.org/10.1002/j.1834-4461.1989.tb00350.x

Evans-Pritchard, E.E. 1937. *Witchcraft, Oracles, and Magic among the Azande*. Oxford: Clarendon Press.

Fabian, Johannes. 1983. *Time and the Other: How Anthropology Makes its Object*. New York: Columbia University Press.

Feng, Xianghong. 2013. "Women's Work, Men's Work: Gender and Tourism among the Miao in Rural China." *Anthropology of Work Review* 34(1):2–14. https://doi.org/10.1111/awr.12002

Ferguson, Lucy. 2010a. "Interrogating 'Gender' in Development Policy and Practice: The World Bank, Tourism and Microenterprise in Honduras." *International Feminist Journal of Politics* 12(1): 3–24. https://doi.org/10.1080/14616740903429080

———. 2010b. "Tourism Development and the Restructuring of Social Reproduction in Central America." *Review of International Political Economy* 17:860–88. https://doi.org/10.1080/09692290903507219

———. 2011. "Tourism, Consumption and Inequality in Central America. *New Political Economy* 16(3):347–71. https://doi.org/10.1080/13563467.2010.500721

Fetterman, David M. 2010. *Ethnography: Step-by-Step*. 3rd ed. Applied Social Research Methods Series No. 17. Los Angeles: Sage.

Foster, George M. 1965. "Peasant Society and the Image of Limited Good." *American Anthropologist* 67(2): 293–315. https://doi.org/10.1525/aa.1965.67.2.02a00010

Fotiou, Evgenia. 2016. "The Globalization of Ayahuasca Shamanism and the Erasure of Indigenous Shamanism." *Anthropology of Consciousness* 27(2):151–79. https://doi.org/10.1111/anoc.12056

Foucault, Michel. 1972. *The Archaeology of Knowledge*. London: Tavistock Publications.

———. 1978. *The History of Sexuality: An Introduction*. vol. 1. New York: Vintage Books.

———. 1980. "Prison Talk." In *Power/Knowledge: Selected Interviews and Other Writings, 1972–1977*, edited by Colin Gordon, 37–54. New York: Pantheon Books.

Franklin, Adrian. 2004. "Tourism as an Ordering: Towards a New Ontology of Tourism." *Tourist Studies* 4(3):277–301. https://doi.org/10.1177/1468797604057328

Frenzel, Fabian, and Ko Koens. 2012. "Slum Tourism: Developments in a Young Field of Interdisciplinary Tourism Research." *Tourism*

Geographies 14(2):195–212. https://doi.org/10.1080/14616688.2012.633222

Friedl, Ernestine. 1991. "Society and Sex Roles." In *Annual Editions: Anthropology 91/92*, edited by Elvio Angeloni, 112–16. Guildford: Dushkin Publishing.

Frohlick, Susan. 2010. "The Sex of Tourism?: Bodies under Suspicion in Paradise." In *Thinking through Tourism*, edited by Julie Scott and Tom Selwyn, 51–70. Oxford: Berg.

———. 2016. "Feeling Sexual Transgression: Subjectivity, Bodily Experience, and Non-Normative Hetero-Erotic Practices in Women's Cross-Border Sex in Costa Rica." *Gender, Place, and Culture* 23(2):257–73. https://doi.org/10.1080/0966369X.2014.991696

Garcia, Alma M., ed. 1997. *Chicana Feminist Thought: The Basic Historical Writings*. vol. 1. New York: Routledge.

Gascón, Jordi. 2016. "Residential Tourism and Depeasantisation in the Ecuadorian Andes." *The Journal of Peasant Studies* 43(4):868–85. https://doi.org/10.1080/03066150.2015.1052964

Geertz, Clifford. 1973. *The Interpretation of Cultures: Selected Essays*. New York: Basic Books.

Ghasarian, Christian. 2014. "Journeys to the Inner Self: Neo-Shamanism and the Search for Authenticity in Contemporary New Age Travel Practice." In *Tourism and the Power of Otherness: Seductions of Difference*, edited by David Picard and Michael A. Di Giovine, 176–91. Bristol: Channel View Publications.

Glidden, Lisa M. 2011. *Mobilizing Ethnic Identity in the Andes*. Lanham: Lexington Books.

Gmelch, Sharon Bohn, ed. 2010. Why Tourism Matters. In *Tourists and Tourism: A Reader*, 3–24. Long Grove: Waveland Press.

Gómez Alcaraz, Fredy Hernán, and Carlos Iván García Suárez. 2006. "Masculinity and Violence in Colombia: Deconstructing the Conventional Way of Becoming a Man." In *The Other Half of Gender: Men's Issues in Development*, edited by Ian Bannon and Maria C. Correia, 93–110. Washington: World Bank.

Gómez-Barris, Macarena. 2012. "Andean Translations: New Age Tourism and Cultural Exchange in the Sacred Valley, Peru." *Latin American Perspectives* 39(6):68–78. https://doi.org/10.1177/0094582X12454561

Graburn, Nelson H.H. 1989. "Tourism: The Sacred Journey." In *Hosts and Guests: The Anthropology of Tourism*, 2nd ed., edited by Valene Smith, 21–36. Philadelphia: University of Pennsylvania Press.

Granholm, Kennet. 2013. "Esoteric Currents as Discursive Complexes." *Religion* 43(1):46–69. https://doi.org/10.1080/0048721X.2013.742741

Greenway, Christine. 1998. "Hungry Earth and Vengeful Stars: Soul Loss and Identity in the Peruvian Andes." *Social Science & Medicine* 47(8):993–1004. https://doi.org/10.1016/S0277-9536(98)00163-4

Greenwood, Davydd. 1989. "Culture by the Pound: An Anthropological Perspective on Tourism as Cultural Commoditization." In *Hosts and Guests: The Anthropology of Tourism*, 2nd ed., edited by Valene L. Smith, 171–85. Philadelphia: University of Pennsylvania Press.

Guelke, Karoline. 2014/2015. "Tourist Photography in Peru: An Actor-Network-Theory Approach to Tourist Photographs Posted Online." *Material Culture Review* 80/81 (Fall/Spring):134–50. https://journals.lib.unb.ca/index.php/MCR/article/view/25561/29637

Guevara Paredes, Mario. 1998. *Gringa Hunter & Other Short Stories*. Cusco: Sieteculebras Editores.

Gutman, Matthew. 1998. "El Machismo." In *Masculinidades y equidad de género en América Latina*, edited by Teresa Valdés and José Olavarría, 238–57. Santiago: Facultad Latinoamericana de Ciencias Sociales.

Hall, Michael C., and Hazel Tucker, eds. 2004. "Tourism and Postcolonialism: An Introduction." In *Tourism and Postcolonialism: Contested Discourses, Identities, and Representations*, 1–24. London: Routledge.

Haraway, Donna J. 1991. *Simians, Cyborgs, and Women: The Reinvention of Nature*. New York: Routledge.

Harper, Douglas. 2002. "Talking about Pictures: A Case for Photo Elicitation." *Visual Studies* 17(1):13–26. https://doi.org/10.1080/14725860220137345

Hayns, Joe. 2016. "Desperate Markets and Desperate Masculinities in Morocco." In *Masculinities under Neocolonialism*, edited by Andrea Cornwall, Frank G. Karioris, and Nancy Lindisfarne, 99–110. London: Zed Books.

Hendrickson, Carol. 2008. "Visual Field Notes: Drawing Insights in the Yucatan." *Visual Anthropology Review* 4(2):117–32. https://doi.org/10.1111/j.1548-7458.2008.00009.x

Henrici, Jane. 2007. "Género, Turismo y Exportación: ¿Llamando a la Plata en el Perú?" *Anthropologica* 25(25):83–101. http://revistas.pucp.edu.pe/index.php/anthropologica/article/view/1425

Herold, Edward, Rafael Garcia, and Tony DeMoya. 2001. "Female Tourists and Beach Boys: Romance or Sex Tourism?" *Annals of Tourism Research* 28(4):978–97. https://doi.org/10.1016/S0160-7383(01)00003-2

Heron, Barbara. 2007. *Desire for Development: Whiteness, Gender, and the Helping Imperative*. Waterloo: Wilfrid Laurier University Press.

Hill, Michael D. 2007. "Contesting Patrimony: Cusco's Mystical Tourist Industry and the Politics of *Incanismo*." *Ethnos* 72(4):433–60. https://doi.org/10.1080/00141840701768276

———. 2008. "Inca of the Blood, Inca of the Soul: Embodiment, Emotion, and Racialization in the Peruvian Mystical Tourist Industry." *Journal of the American Academy of Religion* 76(2):251–79. https://doi.org/10.1093/jaarel/lfn007

Himpele, Jeffrey, and Quetzil Castañeda, directors. 1997. *Incidents of Travel in Chichén Itzá*. Waterton: Documentary Educational Resources.

Hollinshead, Keith. 2004. "Tourism and New Sense: Worldmaking and the Enunciative Value of Tourism." In *Tourism and Postcolonialism: Contested Discourses, Identities and Representations*, edited by Michael C. Hall and Hazel Tucker, 23–42. London: Routledge.

Hubbard, Ethan. 1990. *Journey to Ollantaytambo*. Post Mills: Chelsea Green Publishing Company.

Index Mundi. 2020. Travel & Tourism: Peru. https://www.indexmundi.com/facts/peru/international-tourism

INEI (Instituto Nacional de Estadística e Informática). 2014. Departamento Cusco: Población Total Proyectada y Ubicación. In Directorio Nacional de Municipalidades Provinciales, Distritales y de Centros. https://www.inei.gob.pe/media/MenuRecursivo/publicaciones_digitales/Est/Lib1159/

Ingold, Tim. 2011. *Redrawing Anthropology: Materials, Movements, Lines*. Farnham: Ashgate.

Jordan, Fiona, and Cara Aitchison. 2008. "Tourism and the Sexualisation of the Gaze: Solo Female Tourists' Experiences of Gendered Power, Surveillance and Embodiment." *Leisure Studies* 27(3):329–49. https://doi.org/10.1080/02614360802125080

Kabeer, Naila. 1999. "Resources, Agency, Achievements: Reflections on Measurement of Women's Empowerment." *Development and Change* 30:435–64. https://doi.org/10.1111/1467-7660.00125

Kempadoo, Kamala. 2004. *Sexing the Caribbean: Gender, Race, and Sexual Labour*. New York: Routledge.

Kerstetter, Deborah, and Kelly Bricker. 2009. "Exploring Fijian's Sense of Place after Exposure to Tourism Development." *Journal of Sustainable Tourism* 17(6):691–708. https://doi.org/10.1080/09669580902999196

Kinnaird, Vivian, and Derek Hall. 1996. "Understanding Tourism Processes: A Gender-Aware Framework." *Tourism Management* 17(2): 95–102. https://doi.org/10.1016/0261-5177(95)00112-3

Kok, Ilja, and Willem Timmers, directors. 2012. *Framing the Other*. Breda: Copper Views Film Productions.

Lamphere, Louise. 2009. "The Domestic Sphere of Women and the Public World of Men: The Strength and Limitations of an Anthropological Dichotomy." In *Gender in Cross-cultural Perspective*, 5th ed., edited by Caroline B. Brettell and Carolyn F. Sargent, 90–9. Upper Saddle River: Pearson Prentice Hall.

Little, Walter E. 2008. "Types of Artisanal Production." In *The Oxford Encyclopedia of Women in World History*, vol. 1, edited by Bonnie G. Smith, 159–62. Oxford: Oxford University Press.

Lockwood, Victoria S. 2009. "The Impact of Development on Women: The Interplay of Material Conditions and Gender Ideology." In *Gender in Cross-Cultural Perspective*, 5th ed., edited by Caroline B. Brettell and Carolyn F. Sargent, 510–25. Upper Saddle River: Pearson Prentice Hall.

Lombrana, Laura Millan. 2018. "Peru Protects Rainbow Mountain Tourism from Mining Sector." *Skift*, November 26, 2018. https://skift.com/2018/11/26/peru-protects-rainbow-mountain-tourism-from-mining-sector/

Lugones, María. 2010. "Toward a Decolonial Feminism." *Hypatia* 25(4): 742–59. https://doi.org/10.1111/j.1527-2001.2010.01137.x

Lykes, M. Brinton. 2006. "Creative Arts and Photography in Participatory Action Research in Guatemala." In *Handbook of Action Research*, edited by Peter Reason and Hilary Bradbury, 269–78. London: Sage Publications.

MacCannell, Dean. 1976. *The Tourist: A New Theory of the Leisure Class*. New York: Schocken Books.

———. 1992. *Empty Meeting Grounds: The Tourist Papers*. London: Routledge.

MacDougall, David. 1998. *Transcultural Cinema: Selected Essays*. Princeton: Princeton University Press.

Maoz, Darya. 2006. "The Mutual Gaze." *Annals of Tourism Research* 33(1): 221–39. https://doi.org/10.1016/j.annals.2005.10.010

———. 2008. "The Backpacking Journey of Israeli Women in Mid-Life." In *Backpacker Tourism: Concepts and Profiles*, edited by Kevin Hannam and Irena Ateljevic, 188–98. Clevedon: Channel View Publications.

McCall, Leslie. 2005. "The Complexity of Intersectionality." *Signs: Journal of Women in Culture and Society* 30(3):1771–1800. https://doi.org/10.1086/426800

McCarthy, Carolyn, Karoline A. Miranda, Kevin Raub, Brendan Sainsbury, and Luke Waterson. 2013. *Lonely Planet Peru*. London: Lonely Planet Publications.

McIlwaine, Cathy, and Kavita Datta. 2003. "From Feminising to Engendering Development." *Gender, Place, and Culture* 10(4):369–82. https://doi.org/10.1080/0966369032000155564

Meisch, Lynn. A. 1995. "Gringas and Otavaleños: Changing Tourist Relations." *Annals of Tourism Research* 22(2):441–62. https://doi.org/10.1016/0160-7383(94)00085-9

Merriam-Webster Dictionary. 2018. Gringo. https://www.merriam-webster.com/dictionary. Accessed February 10, 2018.

Mincetur (Ministerio Comercio Exterior y Turismo). 2019. *Perù: Compendio de Cifras de Turismo*. https://www.gob.pe/institucion/mincetur/colecciones/576-peru-compendio-de-cifras-de-turismo

———. 2020. Flujo de Turistas Internacionales e Ingreso de Divisas por Turismo Receptivo. *Datos Turismo: Sistema de Información – Estadística de Turismo*. http://datosturismo.mincetur.gob.pe/appdatosTurismo/Content1.html

Mirandé, Alfredo, and Evangelina Enríquez, 1979. *La Chicana*. Chicago, IL: University of Chicago Press.

Mitchell, Ross E., and Paul F.J. Eagles. 2001. "An Integrative Approach to Tourism: Lessons from the Andes of Peru." *Journal of Sustainable Tourism* 9(1):4–28. https://doi.org/10.1080/09669580108667386

Mitchell, William P. 2006. *Voices from the Global Margin: Confronting Poverty and Inventing New Lives in the Andes.* Austin: University of Texas Press.

Mohanty, Chandra Talpade. 1997. "Under Western Eyes: Feminist Scholarship and Colonial Discourses." In *The Women, Gender and Development Reader*, edited by Nalini Visvanathan, Lynn Duggan, Laurie Nisonoff, and Nan Wiegersma, 79–86. London: Zed Books.

Moore, Henrietta L. 1988. *Feminism and Anthropology.* Oxford: Polity Press.

Mosedale, Sarah. 2005. "Assessing Women's Empowerment: Towards a Conceptual Framework." *Journal of International Development* 7(2):243–57. https://doi.org/10.1002/jid.1212

Murtinho, Felipe, Christina Tague, Bert de Bievre, Hallie Eakin, and David Lopez-Carr. 2013. "Water Scarcity in the Andes: A Comparison of Local Perceptions and Observed Climate, Land Use and Socioeconomic Changes." *Human Ecology* 41(5):667–81. https://doi.org/10.1007/s10745-013-9590-z

Myers, Linda, and Kevin Hannam. 2008. "Women as Backpacker Tourists: A Feminist Analysis of Destination Choice and Social Identities from the UK." In *Backpacker Tourism: Concepts and Profiles*, edited by Kevin Hannam and Irena Ateljevic, 174–87. Clevedon: Channel View Publications.

Nash, Dennison. 1989. "Tourism as a Form of Imperialism." In *Hosts and Guests: The Anthropology of Tourism*, 2nd ed., edited by Valene L. Smith, 37–52. Philadelphia: University of Pennsylvania Press.

Navarro, Marysa. 2002. "Against *Marianismo*." In *Gender's Place: Feminist Anthropologies of Latin America*, edited by Rosario Montoya, Lessie Jo Frazier, and Janise Hurtig, 257–72. New York: Palgrave MacMillan.

Noy, Chaim. 2004. "This Trip Really Changed Me: Backpackers' Narratives of Self-Change." *Annals of Tourism Research* 31(1):78–102. https://doi.org/10.1016/j.annals.2003.08.004

Observatorio National de la Violencia contra las Mujeres y los Integrantes del Grupo Familiar. 2020. *Encuesta Demográphica y de Salud Familiar ENDES – INEI.* https://observatorioviolencia.pe/datos-inei-2017-/#3_Evolucion_de_la_violencia_familiar_en _mujeres_alguna_vez_unidas. Last accessed June 10, 2020.

Olavarría, José. 2006. "Men's Gender Relations, Identity, and Work-Family Balance in Latin America." In *The Other Half of Gender: Men's Issues in Development*, edited by Ian Bannon and Maria C. Correia, 29–42. Washington: World Bank.

Olazabal Castillo, Oscar. 2010. *Ollantaytambo: Cultural, Archaeological, Natural Wonder of Peru.* Lima: Municipalidad Distrital de Ollantaytambo

Orlove, Benjamin. 1998. "Down to Earth: Race and Substance in the Andes." *Bulletin of Latin American Research* 17(2):207–22. https://doi.org/10.1111/j.1470-9856.1998.tb00172.x

O'Rourke, Dennis, director. 1988. *Cannibal Tours.* Australia: Dennis O'Rourke and Laurence J. Henderson.

Oxfam. 2017. "Inequality in Peru: Reality and Risks." https://peru.oxfam.org/sites/peru.oxfam.org/files/file_attachments/Inequality%20in%20Peru.%20Reality%20and%20Risks.pdf

Parisi, Laura. 2013. "Gender Mainstreaming Human Rights: A Progressive Path for Gender Equality?" In *Human Rights: The Hard Questions*, edited by Cindy Holder and David Reidy, 436–57. New York: Cambridge University Press.

Paz y Luz (Hotel & Healing Center). 2017. http://www.pazyluzperu.com/wp/

PhotoVoice. 2020. PhotoVoice: Participatory Photography for Social Change. https://photovoice.org

Pink, Sarah. 2004. "Introduction: Situating Visual Research." In *Working Images: Visual Research and Representation in Ethnography*, edited by Sarah Pink, Laszlo Kürti, Ana Isabel Afonso, 1–12. London: Routledge.

———. 2007. *Doing Visual Ethnography*. London: Sage Publications.

Poole, Deborah. 1997. *Vision, Race, and Modernity: A Visual Economy of the Andean Image World*. Princeton: Princeton University Press.

Pratt, Mary Louise. 1992. *Imperial Eyes: Travel Writing and Transculturation*. London: Routledge.

Prayag, Girish, Paolo Mura, Colin Michael Hall, and Julien Fontaine. 2016. "Spirituality, Drugs, and Tourism: Tourists' and Shamans' Experiences of Ayahuasca in Iquitos, Peru." *Tourism Recreation Research* 41(3):314–25. https://doi.org/10.1080/02508281.2016.1192237

PROMPERÙ. 2020. *Tips Perfil del Turista Extranjero 2019*. https://www.promperu.gob.pe/TurismoIN/sitio/PerfTuristaExt

Pruitt, Deborah, and Suzanne LaFont. 1995. "For Love and Money: Romance Tourism in Jamaica." *Annals of Tourism Research* 22(2):422–40. https://doi.org/10.1016/0160-7383(94)00084-0

Quisumbing, Agnes R. 2010. "Gender and Household Decision-Making in Developing Countries: A Review of Evidence." In *The International Handbook of Gender and Poverty: Concepts, Research, Policy*, edited by Sylvia Chant, 161–66. Cheltenham: Edward Elgar.

Rai, Shirin. 2011a. "The History of International Development: Concepts and Contexts." In *The Women, Gender, and Development Reader*, 2nd ed., edited by Nalini Visvanathan, Lynn Duggan, Nan Wiegersma, and Laurie Nisonoff, 14–21. London: Zed Books.

———. 2011b. "Gender and Development: Theoretical Perspectives. In *The Women, Gender, and Development Reader*, 2nd ed., edited by Nalini Visvanathan, Lynn Duggan, Nan Wiegersma, and Laurie Nisonoff, 28–37. London: Zed Books.

Ramos, Manuel João. 2004. "Drawing the Lines: The Limitations of Intercultural *Ekphrasis*." In *Working Images: Visual Research and Representation in Ethnography*, edited by Sarah Pink, Laszlo Kürti, Ana Isabel Afonso, 147–57. London: Routledge.

Rathgeber, Eva. 1990. "WID, WAD, GAD: Trends in Research and Practice." *Journal of Developing Areas* 24(4):489–505. www.jstor.org/stable/4191904

Rodes, Jennifer, director. 1992. *Trekking on Tradition*. Watertown: Documentary Educational Resources.

Rosaldo, Michelle Z. 1974. "Woman, Culture, and Society: A Theoretical Overview." In *Woman, Culture, and Society*, edited by Michelle Z. Rosaldo and Loise Lamphere, 17–43. Stanford: Stanford University Press.

Rosaldo, Renato. 1989. "Imperialist Nostalgia." *Representations, Special Issue: Memory and Counter-Memory* 26:107–22. https://doi.org/10.2307/2928525

Salazar, Noel B. 2012. "Tourist Imaginaries: A Conceptual Approach." *Annals of Tourism Research* 39(2):863–82. https://doi.org/10.1016/j.annals.2011.10.004

Salazar, Noel B., and Nelson H. H. Graburn, eds. 2014. "Introduction. Toward an Anthropology of Tourism Imaginaries." In *Tourism Imaginaries: Anthropological Approaches*, 1–28. New York: Berghan.

Sammells, Clare A. 2014. "Bargaining under Thatch Roofs: Tourism and the Allure of Poverty in Highland Bolivia." In *Tourism and the Power of Otherness: Seductions of Difference*, edited by David Picard and Michael A. Di Giovine, 124–37. Bristol: Channel View Publications.

Sariego López, Ignacio, and Alberto Moreno Melgarejo. 2011. *Plan Estratégico de Desarrollo Turístico del Distrito de Ollantaytambo al 2015*. Urubamba: Municipalidad Distrital de Ollantaytambo.

Scarles, Caroline. 2010. "Where Words Fail, Visuals Ignite: Opportunities for Visual Autoethnography in Tourism Research." *Annals of Tourism Research* 37(4):905–26. https://doi.org/10.1016/j.annals.2010.02.001

Shields, Stephanie A. 2008. "Gender: An Intersectionality Perspective." *Sex Roles* 59:301–11. https://doi.org/10.1007/s11199-008-9501-8

Silverblatt, Irene. 1983. "The Evolution of Witchcraft and the Meaning of Healing in Colonial Andean Society." *Culture, Medicine and Psychiatry* 7:413–27. https://doi.org/10.1007/BF00052240

———. 1987. *Moon, Sun, and Witches: Gender Ideologies and Class in Inca and Colonial Peru*. Princeton: Princeton University Press.

Silverman, Helaine. 2002. "Touring Ancient Times: The Present and Presented Past in Contemporary Peru." *American Anthropologist* 104(3).881–902. https://doi.org/10.1525/aa.2002.104.3.881

Sinclair, Thea M., ed. 1997. "Issues and Theories of Gender and Work in Tourism." In *Gender, Work, and Tourism*, 1–15. London: Routledge.

Sinervo, Aviva, and Michael D. Hill. 2011. "The Visual Economy of Andean Childhood Poverty: Interpreting Postcards in Cusco, Peru." *Journal of Latin American and Caribbean Anthropology* 16(1):114–42. https://doi.org/10.1111/j.1935-4940.2011.01127.x

Singh, Sagar. 2019. *Rethinking the Anthropology of Love and Tourism*. Lanham: Lexington Books.

Smith, Valene L., ed. 1989. "Introduction." In *Hosts and Guests: The Anthropology of Tourism*, 2nd ed., 1–17. Philadelphia: University of Pennsylvania Press.

Sontag, Susan. 1977. *On Photography*. London: Allan Lane.

Steel, Griet. 2013. "Mining and Tourism: Urban Transformations in the Intermediate Cities of Cajamarca and Cusco, Peru." *Latin American Perspectives* 40(2):237–49. https://doi.org/10.1177/0094582X12468866

Stevens, Evelyn. 1973. "*Marianismo*, the Other Face of *Machismo*." In *Female and Male in Latin America*, edited by Ann Pescatello, 89–101. Pittsburg: University of Pittsburg Press.

Stronza, Amanda. 2001. "Anthropology of Tourism: Forging New Ground for Ecotourism and Other Alternatives." *Annual Review of Anthropology* 30:261–83. https://doi.org/10.1146/annurev.anthro.30.1.261

Swain, Margaret. 1993. "Women Producers of Ethnic Arts." *Annals of Tourism Research* 20:32–51. https://doi.org/10.1016/0160-7383(93)90110-O

Sweet, Jill D. 1989. "Burlesquing 'the Other' in Pueblo Performance." *Annals of Tourism Research* 16:62–75. https://doi.org/10.1016/0160-7383(89)90030-3

———. 2010. "'Let'em Loose': Pueblo Indian Management of Tourism." In *Tourists and Tourism: A Reader*, edited by Sharon Bohn Gmelch, 137–50. Long Grove, IL: Waveland Press.

Taussig, Michael T. 1980. *The Devil and Commodity Fetishism in South America*. Chapel Hill: University of North Carolina Press.

———. 2011. *I Swear I Saw This: Drawings in Fieldwork Notebooks, Namely My Own*. Chicago: University of Chicago Press.

Theodossopoulos, Dimitrios. 2014. "Scorn or Idealization? Tourism Imaginaries, Exoticization, and Ambivalence in Emberá Indigenous Tourism." In *Tourism Imaginaries: Anthropological Approaches*, edited by Noel B. Salazar and Nelson H. H. Graburn, 57–79. New York: Berghan.

———. 2017. "Solidarity: A Graphic Ethnography." Centre for Imaginative Ethnography. http://imaginativeethnography.org/imaginings/comics/solidarity-a-graphic-ethnography/

Tucker, Hazel. 2003. *Living with Tourism: Negotiating Identities in a Turkish Village*. London: Routledge.

Turner, Terence. 1992. "Defiant Images: The Kayapo Appropriation of Video." *Anthropology Today* 8(6):5–16. https://doi.org/10.2307/2783265

Turton, David. 2004. "Lip-Plates and 'the People Who Take Photographs.'" *Anthropology Today* 30(3):3–9. https://doi.org/10.1111/j.0268-540X.2004.00266.x

United States Institute of Peace. 2018. Truth Commission: Peru 01. https://www.usip.org/publications/2001/07/truth-commission-peru-01

UNWTO (United Nations World Travel Organization). 2019. *Global Report on Women in Tourism*. 2nd edition. https://www.unwto.org/publication/global-report-women-tourism-2-edition

———. 2020a. *Global Guidelines to Restart Tourism.* https://webunwto.s3.eu-west-1.amazonaws.com/s3fs-public/2020-05/UNWTO-Global-Guidelines-to-Restart-Tourism.pdf
———. 2020b. "Tourism Can Be a Platform for Overcoming the Pandemic. By bringing people Together, Tourism Can Promote Solidarity and Trust: UN Secretary-General Antonio Guterres." *UNWTO News*, June 9, 2020. https://www.unwto.org/news/tourism-can-promote-solidarity-un-secretary-general-antonio-guterres
Uriely, Natan. 2005. "The Tourist Experience: Conceptual Developments." *Annals of Tourism Research* 32(1):199–216. https://doi.org/10.1016/j.annals.2004.07.008
Urry, John. 1990. *The Tourist Gaze: Leisure and Travel in Contemporary Societies.* London: Sage Publications.
Urry, John, and Jonas Larsen. 2011. *The Tourist Gaze 3.0.* London: Sage Publications.
van den Berghe, Pierre L. 1994. *The Quest for the Other: Ethnic Tourism in San Cristóbal, Mexico.* Seattle: University of Washington Press.
van den Berghe, Pierre L., and Jorge Flores Ochoa. 2000. "Tourism and Nativistic Ideology in Cusco, Peru." *Annals of Tourism Research* 27(1): 7–26. https://doi.org/10.1016/S0160-7383(99)00043-2
Virmani, Amit, director. 2009. *Cowboys in Paradise.* Singapore: Coup Communications.
Walter, Lynn. 1995. "Feminist Anthropology?" *Gender & Society* 9(3):272–88. https://doi.org/10.1177/089124395009003002
Wang, Caroline, and Mary Ann Burris. 1997. "Photovoice: Concept, Methodology, and Use for Participatory Needs Assessment." *Health Education & Behavior* 24:369–87. https://doi.org/10.1177/109019819702400309
Wang, Caroline C., and Yanique A. Redwood-Jones. 2001. "Photovoice Ethics: Perspectives from Flint Photovoice." *Health Education & Behavior* 28:560–71. https://doi.org/10.1177/109019810102800504
Wang, Ning. 1999. "Rethinking Authenticity in Tourism Experience." *Annals of Tourism Research* 26(2):349–70. https://doi.org/10.1016/S0160-7383(98)00103-0
Waring, Marilyn. 1988. *If Women Counted: A New Feminist Economics.* San Francisco: Harper & Row.
Wehner, Ross, and Renée del Gaudio. 2011. *Moon Handbook: Cusco & Machu Picchu.* Berkeley: Perseus Books Group.
Weismantel, Mary. 1988. *Food, Gender, and Poverty in the Ecuadorian Andes.* Philadelphia, PA: University of Pennsylvania Press.
———. 2001. *Cholas and Pishtacos: Stories of Race and Sex in the Andes.* Chicago: University of Chicago Press.
Welk, Peter. 2004. "The Beaten Track: Anti-Tourism as an Element of Backpacker Identity Construction." In *The Global Nomad: Backpacker Travel*

in Theory and Practice, edited by Greg Richards and Julie Wilson, 77–91. Clevedon: Channel View Publications

Werber, Nick. 2016. *Imasmari.* https://vimeo.com/176568260. Accessed January 10, 2017.

Williams, Erica Lorraine. 2013. *Ambiguous Entanglements: Sex Tourism in Bahia.* Urbana: University of Illinois Press.

Wilson, Erica, and Irena Ateljevic, eds. 2008. "Challenging the 'Tourist-Other' Dualism: Gender, Backpackers, and the Embodiment of Tourism Research." In *Backpacker Tourism: Concepts and Profiles*, 95–110. Clevedon: Channel View Publications.

Wilson, Tamar Diana, and Annelou Ypeij. 2012. "Introduction: Tourism, Gender, and Ethnicity." *Latin American Perspectives* 39(187):5–16. https://doi.org/10.1177/0094582X12453896

Winston, Brian. 2005. "'The Camera Never Lies': The Partiality of Photographic Evidence. In *Image-Based Research: A Sourcebook for Qualitative Researchers*, edited by Jon Prosser, 53–60. Philadelphia: Falmer Press.

World Bank. 2017. Country Poverty Brief: Peru. http://databank.worldbank.org/data/download/poverty/B2A3A7F5-706A-4522-AF99-5B1800FA3357/9FE8B43A-5EAE-4F36-8838-E9F58200CF49/60C691C8-EAD0-47BE-9C8A-B56D672A29F7/Global_POV_SP_CPB_PER.pdf

WTTC (World Travel & Tourism Council). 2020. Travel and Tourism – Economic Impact 2020: Peru. www.wttc.org

York, Michael. 2001. "New Age Commodification and Appropriation of Spirituality." *Journal of Contemporary Religion* 16(3):361–72. https://doi.org/10.1080/13537900120077177

Ypeij, Annelou. 2012. "The Intersection of Gender and Ethnic Identities in the Cuzco-Machu Picchu Tourism Industry: *Sácamefotos*, Tour Guides, and Women Weavers." *Latin American Perspectives* 39(6):17–35. https://doi.org/10.1177/0094582X12454591

Ypeij, Annelou, and Elayne Zorn. 2007. "Taquile: A Peruvian Tourist Island Struggling for Control." *European Review of Latin American and Caribbean Studies* 82:119–28. https://doi.org/10.18352/erlacs.9643

Zainuddin, Anizah H. 2009. "Using Photo Elicitation in Identifying Tourist Motivational Attributes for Visiting Taman Negara, Malaysia." *Management Science and Engineering* 3(1):9–17. http://dx.doi.org/10.3968/j.mse.1913035X20080301.002

Zoomers, Annelies. 2008. "Global Travelling along the Inca Route: Is International Tourism Beneficial for Local Development?" *European Planning Studies* 16(7):971–83. https://doi.org/10.1080/09654310802163769

Zorn, Elayne. 2004. *Weaving a Future: Tourism, Cloth, and Culture on an Andean Island.* Iowa City: University of Iowa Press.

Index

Italicized page numbers indicate figures, maps, or tables.

TC TEACHING CULTURE

Ethnographies for the Classroom

Editor: John Barker, University of British Columbia

This series is an essential resource for instructors searching for ethnographic case studies that are contemporary, engaging, provocative, and created specifically with undergraduate students in mind. Written with clarity and personal warmth, books in the series introduce students to the core methods and orienting frameworks of ethnographic research and provide a compelling entry point to some of the most urgent issues faced by people around the globe today.

Recent Books in the Series

The Living Inca Town: Tourist Encounters in the Peruvian Andes by Karoline Guelke (2021)

Collective Care: Indigenous Motherhood, Family, and HIV/AIDS by Pamela J. Downe (2021)

I Was Never Alone, or Oporniki: An Ethnographic Play on Disability in Russia by Cassandra Hartblay (2020)

Millennial Movements: Positive Social Change in Urban Costa Rica by Karen Stocker (2020)

From Water to Wine: Becoming Middle Class in Angola by Jess Auerbach (2020)

Deeply Rooted in the Present: Heritage, Memory, and Identity in Brazilian Quilombos by Mary Lorena Kenny (2018)

Long Night at the Vepsian Museum: The Forest Folk of Northern Russia and the Struggle for Cultural Survival by Veronica Davidov (2017)

Truth and Indignation: Canada's Truth and Reconciliation Commission on Indian Residential Schools, second edition, by Ronald Niezen (2017)

Merchants in the City of Art: Work, Identity, and Change in a Florentine Neighborhood by Anne Schiller (2016)

Ancestral Lines: The Maisin of Papua New Guinea and the Fate of the Rainforest, second edition, by John Barker (2016)

Love Stories: Language, Private Love, and Public Romance in Georgia by Paul Manning (2015)

Culturing Bioscience: A Case Study in the Anthropology of Science by Udo Krautwurst (2014)

Fields of Play: An Ethnography of Children's Sports by Noel Dyck (2012)